FOREST CONSERVATION POLICY

A Reference Handbook

Other Titles in ABC-CLIO's
**CONTEMPORARY
WORLD ISSUES**
Series

Books in the Contemporary World Issues series address vital issues in today's society such as terrorism, sexual harassment, homelessness, AIDS, gambling, animal rights, and air pollution. Written by professional writers, scholars, and nonacademic experts, these books are authoritative, clearly written, up-to-date, and objective. They provide a good starting point for research by high school and college students, scholars, and general readers as well as by legislators, businesspeople, activists, and others.

Each book, carefully organized and easy to use, contains an overview of the subject; a detailed chronology; biographical sketches; facts and data and/or documents and other primary-source material; a directory of organizations and agencies; annotated lists of print and nonprint resources; a glossary; and an index.

Readers of books in the Contemporary World Issues series will find the information they need in order to better understand the social, political, environmental, and economic issues facing the world today.

FOREST CONSERVATION POLICY

A Reference Handbook

V. Alaric Sample and Antony S. Cheng

CONTEMPORARY WORLD ISSUES

A B C 🍃 C L I O

Santa Barbara, California • Denver, Colorado • Oxford, England

Library of Congress Cataloging-in-Publication Data

Sample, V. Alaric.
  Forest conservation policy : a reference handbook / V. Alaric Sample and Antony S. Cheng.
      p. cm. -- (Contemporary world issues)
  ISBN 1-57607-991-0 (hardcover : alk. paper)
  ISBN 1-57607-992-9 (eBook)
1. Forest conservation--United States--Handbooks, manuals, etc. 2. Forest policy--United States--Handbooks, manuals, etc. I. Cheng, Antony S. II. Title. III. Series.

  SD412.S26 2003
  333.75'16'0973--dc22

                                              2003019575

07 06 05 04   10 9 8 7 6 5 4 3 2 1

This book is also available on the World Wide Web as an eBook. Visit abc-clio.com for details.

ABC-CLIO, Inc.
130 Cremona Drive, P.O. Box 1911
Santa Barbara, California 93116-1911

This book is printed on acid-free paper ∞.

Manufactured in the United States of America

Contents

Preface, xiii

1 **The Conservation Movement and the Roots of U.S. Forest Policy, 1**
Forest Policy in an Era of Environmental Awareness, 1
Historical Roots, 4
Early Forest Laws, 5
Forest Exploitation and Disposal of the Public
      Domain, 6
Forest Reserves and the Awakening of a Conservation
      Ethic, 8
Forest Management Policies in the National Forests, 11
   From Custodial Management to Intensive Use, 11
   Multiple-Use Forestry, 12
   Wilderness, 14
   The Clearcutting Controversy, 15
Establishing Sustainable Forest Management on Private
      Lands, 16
   Evolution of the Forest Products Industry, 17
   Improving Management of Private Forests, 18
Conclusion, 20
References, 22

2 **Current Issues in U.S. Forest Policy, 23**
Sustainable Forest Management, 24
   Background, 25
   Recent Policy Developments, 27
   Conclusion, 32

Conserving Biological Diversity, 33
   Background, 33
   Recent Policy Developments, 35
   Conclusion, 40
Wilderness and Roadless Areas, 41
   Background, 41
   Recent Policy Developments, 44
   Conclusion, 45
Forest Health, 47
   Background, 47
   Recent Policy Developments, 50
   Conclusion, 52
Community Forestry, 53
   Background, 54
   Basic Elements of Community Forestry, 56
   Recent Policy Developments, 58
   Conclusion, 61
Forestry Regulation on Private Lands, 62
   Background, 62
   First Generation: 1920–1950, 62
   Second Generation: 1970–1983, 65
   Third Generation: 1984–1995, 70
   Conclusion, 73
Forest Land Conservation, 74
   Conservation of Public Forest Lands, 75
   Conservation of Corporate Forest Lands, 77
   Conservation of Family Forest Lands, 79
Forest Management Certification, 82
   Background, 83
   Recent Policy Developments, 86
Forests and Water, 89
   Background, 89
   Recent Policy Developments, 91
   Conclusion, 93
Forests and Atmospheric Carbon, 94
   Background, 94
   Recent Policy Developments, 97
   Conclusion, 100
Forest Biotechnology, 101
   Background, 101
   Recent Policy Developments, 103
   Conclusion, 104

Emerging Issues and Future Policy Directions, 106
  Wood Supply in the Global Context, 106
  Forest Management as a Plus in Biodiversity
    Conservation, 108
  Putting U.S. Forest Policy in the Global Context,
    109
References, 110

3 Chronology, 117

4 Personalities, 131
  Bolle, Arnold (1912–1993), 131
  Church, Frank (1924–1984), 131
  Dana, Samuel Trask (1883–1978), 132
  Dwyer, William (1929–2002), 132
  Fernow, Bernhard (1851–1923), 132
  Franklin, Jerry (1936–), 133
  Gordon, John (1939–), 133
  Graves, Henry (1871–1951), 133
  Hough, Franklin (1822–1885), 134
  Humphrey, Hubert (1911–1978), 134
  Ickes, Harold (1874–1952), 134
  Johnson, K. Norman (1942–), 135
  Leopold, Aldo (1886–1948), 135
  Marsh, George Perkins (1801–1882), 135
  Marshall, Robert (1901–1939), 136
  Mather, Stephen (1867–1930), 136
  McGuire, John (1916–2002), 136
  Muir, John (1838–1914), 136
  Murie, Olaus (1889–1963), 137
  Olmsted, Frederick Law (1822–1903), 137
  Pack, Charles Lathrop (1857–1937), 137
  Peterson, Max (1927–), 137
  Pinchot, Gifford (1865–1946), 138
  Powell, John Wesley (1834–1902), 138
  Roosevelt, Theodore (1858–1919), 139
  Sargent, Charles Sprague (1841–1927), 139
  Schenck, Carl (1868–1955), 139
  Thomas, Jack Ward (1934–), 139
  Udall, Morris (1922–1988), 140
  Weyerhaeuser, Frederick (1834–1912), 140
  Weyerhaeuser, Frederick E. (1872–1945), 140

Zahniser, Howard (1906–1964), 141
Zon, Raphael (1887–1957), 141

5 **Forest Facts and Data, 143**
Forest Ecoregions in the United States, 144
Forest Area, Production, and Trade Information, 145
Ownership of U.S. Forests, 146
Federal Public Forest System, 147
    The National Forest System, 148
    Bureau of Land Management, 151
    National Park System, 153
    The National Wilderness Preservation System, 153
State and County Forest Lands, 153
Private Forest Lands, 154
    Forest Industry, 154
    Timber Investment Management Organizations, 155
    Nonindustrial Private Forests, 155
Tribal Forests, 157
Conditions and Trends in U.S. Forests, 158
The Montreal Process, 161
    Criterion 1 through 6, 162
    Criterion 7, 168
Forest Certification Principles and Standards, 171
    SmartWood, 171
    Scientific Certification Systems, 173
    The Sustainable Forestry Initiative, 175
    American Tree Farm System Standards, 176
References, 179

6 **Directory of U.S. Forestry Organizations, 183**
Federal and International Government Organizations,
    183
State Organizations, 190
Nongovernment Conservation Organizations, 197
Society of American Foresters Accredited Forestry
    Schools, 220
Forest Products Companies, 227
Forestry Trade Associations, 229
Forestry Professional Associations, 232
Other Forestry-Related Associations, 234

7   **Print Resources, 239**
General Forestry and Forest Conservation Books and
    Reports, 240
U.S. Forest Policy Books and Reports, 243
U.S. Forest History, 248
International Forestry and Forest Conservation, 250
Scientific Journals, 252

8   **Nonprint Resources, 265**
Information Clearinghouses on the Web, 265
Online Reports, Maps, and Information, 267
Videos, 270
Nonprint Resources from Various States, 273

Glossary, 277
Index, 293
About the Authors, 321

Preface

Through history, forests have played a central role in developing nations, the United States included. The exploitation of these primeval natural resources has built and fueled human habitation since ancient times. Forests have brought wealth and power to nations through trade and conquest. They have even served as collateral in the building of empires, as when Marc Antony made the forests of what is now southern Turkey a birthday gift to Cleopatra, who used them to build a powerful navy. From Odysseus to Horatio Lord Nelson, forests have been key to exploration and trade between nations, and the power of empires has often risen or fallen in step with the health and well-being of their forests.

From a broad perspective, it was not long ago that the United States itself was a young and developing nation, with citizens who were exploiting a continent new to Europeans and clothed with a seemingly inexhaustible supply of trees. Barely a century ago, the primeval forest resources of the United States were being removed at a far higher rate than they could possibly be replaced, and little if any thought was given to their regeneration. Fortunately for this particular developing nation, however, the United States did not continue on this path until its forests were exhausted and ruined. Inspired in part by the history of denuded and barren lands once ruled by the classical empires—as described in 1864 by George Perkins Marsh in his landmark book *Man and Nature*—farsighted and public-spirited individuals including Gifford Pinchot, Bernhard Fernow, and Franklin Hough acted to conserve America's forests to ensure their protection and sustainable use in perpetuity. Now, a century after these and other conservationists established the policies and institutions that halted rampant forest exploitation and developed a new science of sustainable forest management, America's forests are nearly as extensive as they were before European colonization.* Today these policies and institutions are a model for many developing nations that seek to sustain their forests for the wealth of ecological, economic, and social values they provide.

The policies that guide the management of U.S. forests continue to be a source of passionate debate, precisely because of the many different resources and uses that forests provide. Public policy is a manifestation of public values. Forest policy—perhaps more than any other area of public policy in the United States—is based on an amalgam of public values as diverse as the 286 million citizens who make up our constitutional democracy. Citizens are directly involved in setting goals and objectives for forest management on public forest lands and in protecting important environmental values on private forests. The accomplishment of these goals within boundaries established through policy is carefully monitored by citizens' groups, public agencies, and corporations that own and manage forest lands. An added complication is that the objectives for which U.S. forests are managed are not static; they evolve over time as social values themselves evolve. Thus, forest policy is dynamic and ever-changing. The fact that forest management in the United States continues to be a focus of debate is a reflection of the value and importance of forests and a tribute to the citizens who actively participate in the governance of natural resources. These resources, after all, are as central to the continued ecological and economic health of the United States as they were to the less-democratic ancient empires that failed to conserve these essential components of national health and well-being.

The issues that define today's U.S. forest policy debates are

---

* More than three-quarters of the conversion of U.S. forests to agriculture and other uses occurred in the nineteenth century, and today only about 5 percent of the area of native forest ("old growth") remains. Since 1907, however, the total area of forest land in the United States has steadily increased, and forests now cover about 71 percent of the area that was forested in 1630. Today, approximately 52 million acres, or 7 percent of U.S. forest land, is reserved or protected in National Parks and Wilderness areas (Smith et al. 1997).

the core of this book. As one in the ABC-CLIO *Contemporary World Issues* series, this book is intended to be a reference handbook rather than a comprehensive text on forest policy. It will be useful to all students interested in forest policy including upper-level undergraduate and graduate students in forestry, conservation, and environmental policy. It will also be useful to journalists and other readers seeking to better understand current policy issues by seeing them both in their historical context and as they relate to one another. We hope readers will come away with a clearer understanding not only of what the policy issues are, but also of why they are issues and for whom they are issues. For those who wish to delve further into these issues, the book serves as a guide to the broader literature on forest policy and to an array of additional informational resources in both print and electronic form.

We wish to thank Michele Burns, currently a graduate student in the Department of Forest Sciences at Colorado State University, for her assistance in compiling the wealth of information on the array of public, private, and nongovernmental organizations involved in U.S. forestry, forest conservation, and forest policy. The United States is unique in the world in the extent and complexity of such organizations—another reflection of the level of citizen involvement in determining the policies that guide the management of our forests. We are also grateful for Michele's assistance in assembling the list of additional print resources and the vast array of information now available electronically via the Internet.

The editorial staff at ABC-CLIO has been wonderful to work with. Special thanks go to Alicia Merritt, who patiently guided this effort from conception to completion; Deborah Lynes, who managed production; and Linda Bevard, whose careful editing will be appreciated by both the writers and readers of this book.

As urbanized and technology-oriented as the United States has become, our forests continue to be key to the ecological, economic, and social well-being of the nation. With barely a century separating our nation's adolescence and its position today as a leading industrialized nation, it is important that we not lose sight of the importance of our forests as sources of abundant clean water, habitat for a wide diversity of animal and plant species, wood for material and renewable energy needs, and a place of refuge and renewal from the pressures of everyday life.

For the United States perhaps more than for any other nation, forests are a part of our history and culture and help to define who we are as a people. We hope this book will be one more small contribution to the citizens of all kinds who continue to play an active role in determining the policies that guide the conservation and sustainable management of our forests.

V. Alaric Sample
Washington, D.C.

Antony S. Cheng
Fort Collins, Colorado

1

The Conservation Movement and the Roots of U.S. Forest Policy

Forest Policy in an Era of Environmental Awareness

At no time in history has there been greater interest and concern over the future of the world's forests. Images from satellites orbiting the Earth convey to us how finite our resources are, and how small a proportion of our sky-blue planet is forest-green. From this satellite imagery we can also see continental-scale smoke plumes from fires that annually deforest areas of the tropics as large as entire U.S. states.

Down here on the ground, some of the most respected scientists of our time tell us that plant and animal species, many of them dependent on forest ecosystems, are quietly going extinct—at a rate faster than at any time since a comet smashed into the Earth 60 million years ago and wiped out the dinosaurs. As 20,000-year-old glaciers melt in Greenland and Antarctica, scientists tell us that the long-term survival of our own species may be in jeopardy from global climate changes leading to rising sea levels, prolonged droughts in key agricultural regions, and more intensely destructive storms. Although global warming is discussed primarily in terms of carbon dioxide and other "greenhouse gases" produced by the combustion of oil, gas, and other

1

fossil fuels, it is estimated that one-quarter of all greenhouse gas emissions stemming from human activity during the past two decades has come from deforestation in the tropics.

Is the situation hopeless? No. But we will need to follow a different course than the one we have been on if this trend of species extinction and global warming is to be changed. Approaches to forest conservation and sustainable use that worked well a half-century ago are unlikely to work a half-century in the future. World population will have more than doubled in that time, and with it the needs for all the goods and services that forests provide. Just as importantly, there is likely to be a similar increase in the needs for agricultural production, living space, and other human activities that displace forests and lead to continued net decreases in forest area.

Forest policy provides a framework for effective action. Policy change comes about when enough people come to regard the current situation or direction as unacceptable and begin to search for alternatives. Policy change is a process by which diverse interests within society arrive at a common set of goals and then organize institutions and legal frameworks that will lead to the achievement of those goals. Forest policy does not stay the same for long. The past few decades have seen a revolution in our scientific understanding of forests—how they function, what they provide, the ways in which they become degraded, and what is needed to sustain them. It is natural that forest policy should continue to evolve to reflect new knowledge and changing social values.

The evolution in U.S. forest policy has in many ways been a prelude to the policy debates now taking place at the global level. A new environmental awareness arose in the United States beginning in the 1960s. Emblematic of this period was the 1962 publication of Rachel Carson's *Silent Spring,* which documented the lethal effects of pesticides and other chemical toxins on songbirds and other species. Carson's book also helped usher in a new era of social activism on behalf of the natural environment, with individual citizens taking to the street and to the halls of government, forcing industries of all kinds to reduce pollution and environmental degradation.

This activism produced several landmark U.S. laws and government policies during the 1970s (among them the National Environmental Policy Act of 1970, Clean Water Act Amendments of 1972 and 1977, and Clean Air Act Amendments of 1970) that provided an entirely new framework for reducing the negative effects of human activities on the air, water, and land on which we all depend. At times, government itself became the target of this new environmental activism. Federal and state agencies created to implement environmental laws were challenged in court when they were seen to be lax in their enforcement of these laws. Other government agencies, particularly those responsible for managing public lands and natural resources, were challenged as never before to demonstrate that they themselves were not contributing to the degradation of important environmental values.

Forests and the policies governing them became an especially high-profile focus of this environmental activism. Concerns over clearcutting and other aspects of commercial timber production prompted a series of Federal and state laws aimed at limiting or prohibiting controversial forest practices, protecting wilderness areas, and conserving habitat for wildlife—especially threatened or endangered species.

Although many of us tend to think of these issues as they relate to our own locale and experiences, they are but a reflection of the broader concern over these very same issues worldwide. For example, the creation of protected forest areas to conserve biological diversity is an issue in every country that has significant areas of forest, whether tropical, temperate, or boreal. The Convention on Biological Diversity, an agreement developed at the United Nations Conference on Environment and Development (UNCED) in 1992, was one of the first major efforts by nations around the world to reach a common understanding and develop a common strategy to conserve biological diversity in the world's forests.

Likewise, our struggle to develop solutions to these issues is a reflection of similar efforts worldwide. Issues in the United States in the 1970s over clearcutting and the protection of environmental values in forests were but a precursor to the current global debate over sustainable forest management. As in the United States, the global debate has been like a pendulum, swinging the focus from commercial uses to environmental values and finally to a position recognizing that, in the long run, forests can only be conserved through a strategy that protects forest productivity and environmental values, maintains economic values, and deals squarely with perennial concerns over the fair and equitable use of these important resources.

Historical Roots

Forest policy is the framework of laws, regulations, and decisions governing the management and use of a variety of forest resources so as to meet current needs while maintaining the value and productivity of forests for future generations. Declared forest policies in the United States date back to the earliest days of European colonization. Native Americans had their own style of forest policies prior to European colonization. Evidence of fairly widespread intentional burning suggests that many Native Americans manipulated the forest environment to increase forage for game and protect themselves from enemies (Pyne 1982; Williams 1989). Alteration of forests became far more widespread and enduring in the years following settlement by European colonists in the early 1600s. Even in those early years, competition over forest resources prompted the sense of resource scarcity and need for regulated use that underlies much of U.S. forest policy today.

Today's forest policies and much of U.S. law in general have their roots in English common law (Adams 1993). Oppressive forest laws in Anglo-Saxon and Norman England, which often mandated punishment by mutilation or death for poaching the king's game or cutting down trees in the royal forests (Young 1979), helped create the impetus for the Magna Carta, the greatest value of which to succeeding generations has been in defining and protecting individual rights. Forest policy today continues to be as much about promoting economic opportunity and ensuring social equity as it is about ecological science.

The balancing of individual and community rights is at the heart of the U.S. Constitution and Bill of Rights, and nowhere in U.S. law is this principle more evident than in the policies governing the conservation and use of forest resources on public and private forest lands. At the start of the twentieth century, forester, conservationist, and social activist Gifford Pinchot wrote that the nation's forests should be managed "for the greatest good, for the greatest number, in the long run" (Pinchot 1911). In a reflection of the defining goals of the Progressive Era, Pinchot worked to create forest policies to eliminate waste of public resources, combat government corruption and corporate fraud, and ensure social equity. A century later, national and international protocols are being developed to achieve "sustainable forestry," which is defined as forest practices that are "ecologically sound, economically viable and socially responsible" (Aplet et al. 1993). It is clear that U.S. forest policy is continuously evolving to reflect not only the changing conditions in forests and their ecological environment but also the changing values and perceptions in the social and economic environment. At their heart, U.S. forest policies are expressions of social values blended with ecological realities.

Following is a brief historical overview of the development of forest policies in the United States. References are provided at the end of this chapter to facilitate a more in-depth examination of the roots of today's forest policy efforts. Chapter 2 more closely examines a few of the key issues that are being addressed in current debates and that will shape the evolution of U.S. forest policy.

Early Forest Laws

Forest policies arose on the North American continent shortly after the arrival of the first Europeans. Less than five years after the *Mayflower* first anchored in Massachusetts Bay in 1620, the Plymouth Colony was already concerned about timber supplies and had enacted a law forbidding "the sale or transport of any timber whatsoever out of the colony without the approval of the governor and the council" (Dana 1956). In 1691, "all pine trees fit for masts" were reserved to the King of England and marked with the "Broad Arrow" symbol of the British Navy. Colonists cutting such trees could be fined or imprisoned. So unpopular was this forest policy that resistance was widespread, and the Broad Arrow policy was one of the irritants that led to the Declaration of Independence and the American Revolution.

Following the Declaration of Independence, the 13 original states took claim to all the lands formerly held by the British Crown. Virginia, North Carolina, South Carolina, Georgia, New York, and Connecticut all claimed extensive "western reserve" lands—Connecticut's western reserve claim, for example, included what is now Chicago. After the Revolutionary War, these states were persuaded to give up their claims, and in 1780 the Continental Congress determined that all western reserve lands ceded to the central government of the United States would be used for the common benefit of all the states. After all, the infant U.S. Treasury was severely depleted from fighting the British, and land was one of the few assets on which the Federal government could draw. The states' shared stake in the economic benefits to be derived from these commonly held lands helped cement the early Union. These state cessions, estimated at about 233 million acres, formed the nucleus of what would become known as the "public domain" of Federal public lands. In the ensuing years, this public domain grew significantly through land purchases and treaties such as the Louisiana Purchase (1803), the Florida Purchase (1819), the Oregon Compromise (1846), the Mexican Cession (1848), the Texas Purchase (1850), the Gadsden Purchase (1853), and the Alaska Purchase (1867), until the public domain totaled more than 1.8 billion acres.

Forest Exploitation and Disposal of the Public Domain

Originally, it was not intended that the public domain lands would remain in federal ownership. Rather, title to these lands was to be conveyed to farmers, settlers, and other private owners as quickly as possible. So began the era of disposal of public domain lands to private citizens. This process began as soon as the Revolutionary War ended, with regular soldiers being granted title to tracts of public domain lands as a bounty for their service. The General Land Ordinance of 1785 provided for the purchase of public domain lands for as little as $1 per acre, and as much as 20 million acres per year were being sold by 1836. Together with the Northwest Ordinance of 1787, the General Land Ordinance was the means by which the public domain was surveyed, divided into townships and sections, and distributed into private hands. The legacy of townships and sections is evident even today on every U.S. Geological Survey quadrangle map that delineates township and section lines. The disposal of the public domain lands was greatly accelerated by the passage of the Homestead Act in 1862, which allowed individuals to obtain up to 160 acres free of charge in return for proof that they had resided on and cultivated the land for five years. About 287 million acres of the public domain were put into private hands under the Homestead Act. In an attempt to address the chronic shortage of timber for developing farms and towns in the Midwest and Plains states, Congress in 1873 passed the Timber Culture Act, authorizing grants of up to 160 acres to anyone willing to plant 40 of these acres with trees and keep the trees healthy and growing for ten years.

In addition, enormous areas of public-domain lands were given to large companies, most notably to the railroads, as a way of expediting the settlement of the western states. Transcontinental grants were made in 1862 to the Union Pacific and Central Pacific Railroads, giving title to alternating 640-acre tracts of Federal land for 20 miles on each side of the railroad over its entire length. Eventually totaling more than 94 million acres, these grants remain the largest conveyances of public land to private corporations in the history of the nation.

Timber trespass—that is, the illegal removal of commercial timber from public lands—was rampant during this period. Rarely was the law effectively enforced and only then with great difficulty and danger. In one instance in 1853, a violent riot had to be put down by armed Marines from the USS *Michigan* after Federal agents tried to seize a large volume of timber illegally cut on public lands near Manistee, Michigan, and to arrest the perpetrators. To help prevent such confrontations, Congress enacted provisions by which timber could legally be obtained from public lands. The Free Timber Act of 1878 allowed citizens in the Rocky Mountain states to obtain free of charge whatever timber they needed for building, agricultural, mining, or other domestic purposes. The Timber and Stone Act of 1878 allowed citizens in Oregon, Washington, California, and Nevada to purchase the land itself for as little as $2.50 per acre to satisfy their needs for building materials (USDA, U.S. Forest Service 1993).

These early forest policies, though well-intentioned, became the basis for monumental fraud, collusion, graft, and outright theft of vast amounts of public land and timber. Schemes breathtaking in their magnitude, boldness, and ingenuity eventually resulted in many millions of acres of some of the most productive forest lands in the West being fraudulently obtained by the largest timber operators of the time—at little or no cost to the companies but at great expense to the public domain and the national interest. Between 1800 and 1910, some 300 million acres, or 469,000 square miles, of forests were cut and converted to nonforest land uses—about 13 percent of the total land area of the United States and nearly one-quarter of the original total forest land area in the United States. The period between 1850 and 1880 was especially devastating; in this 30-year period, nearly 100 million acres were deforested—an area nearly twice the size of Kansas.

Forest Reserves and the Awakening of a Conservation Ethic

In the history of forest policy in the United States, the Forest Reserve Act of 1891 marks a major turning point. The overexploitation of forests on private lands and the resulting local timber shortages, destructive wildfires, costly floods, and sharp decrease in game populations had persuaded early conservationists that it made sense to reserve a significant area of the remaining public domain lands in federal ownership, to be protected and managed for the benefit of the entire nation rather than a privileged few.

One of the earliest cautionary tales about the danger of forest destruction was *Man and Nature,* written in 1864 by George Perkins Marsh (Lowenthal 1965). Marsh recounts the fall of great civilizations, from Babylonia to the Roman Empire, in part because they overexploited their forests. By the 1870s, it was becoming clear that forests in the United States were not as inexhaustible as previously thought and that the nation's timber supply was being drawn down far more quickly than it was being replenished. America was still a nation that ran on wood. Almost all buildings were constructed of wood; all major forms of transportation relied on wood (for example, sailing ships, railroads, bridges, carriages); and wood was the predominant fuel for heating, cooking, and many industrial processes. The prospect of a "timber famine" was grim indeed, and guaranteeing a timber supply adequate for the needs of a growing nation became the primary driver of U.S. forest policy for most of the century to follow. Groups such as the American Forestry Association and the American Association for the Advancement of Science pushed hard for Congress to take action.

The resulting Forest Reserve Act of 1891 authorized the President of the United States to reserve forested lands from the public domain and discontinued virtually all disposal of public forest lands into private ownership. By 1897, almost 40 million acres had been proclaimed in more than a dozen Forest Reserves created throughout the western United States. But because the law provided no guidance on how the reserves were to be used and managed, western states and territories clamored that the resources had simply been "locked up." To alleviate this situation, Congress that year passed the Organic Administration Act, specifying the purposes for which Forest Reserves could be designated and how they were to be managed. Forest Reserves were to be established "to improve and protect the forest within the reservation for the purpose of securing favorable conditions of water flows, and to furnish a continuous supply of timber for the use and necessities of citizens of the United States." It instructed the Secretary of the Interior to protect the Forest Reserves from fire and theft and authorized the Secretary to "make such rules and regulations and establish such service as will insure the objects of such reservations, namely, to regulate their occupancy and use and to preserve the forests thereon from destruction."

By 1905, management of the Forest Reserves—they were renamed *National Forests* in 1907—was transferred to the U.S. Forest Service, an agency within the U.S. Department of Agriculture. Under the charismatic leadership of Gifford Pinchot, its first Chief Forester, the Forest Service quickly distinguished itself as a highly decentralized agency of professionally trained foresters who were capable of making technically proficient decisions at the local level. Pinchot was a strong advocate of developing a cadre of professionals insulated from political pressures. The creation of the U.S. Forest Service as an independent civil service agency was consistent with the ideas of the Progressive Movement, of which Pinchot was a strong proponent. The Progressives' approach to forest and rangeland management focused not on strict protection but on regulating extractive uses such as timber cutting or livestock grazing to levels that could be sustained over the long term and that would protect public values such as water, wildlife, and soil productivity. Later these values would include recreation and wilderness. Between 1906 and 1909, President Theodore Roosevelt, a close friend and confidant of Gifford Pinchot, set aside an additional 80 million acres of National Forests.

In 1911, Congress authorized the Forest Service to purchase private lands for the creation of new National Forests. Up until this time, the National Forests had all been created out of public domain lands by Presidential authorization, generally west of the Mississippi River. In the early 1900s, there were large areas of land in the eastern forests that had been privately owned, but they had been stripped of their timber and abandoned. Many of these lands had become tax-delinquent and reverted to state or county ownership. Under the Weeks Act of 1911, the Forest Service acquired many thousands of acres of such land, protected it from fire, and gradually restored the extensive, mature forests that now exist in National Forests from the Appalachians west to the Great Plains. (Figure 1.1 shows the National Forest System today.)

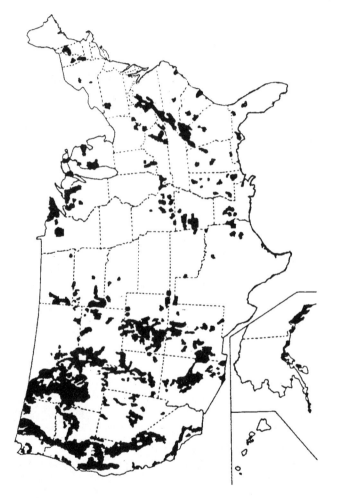

Figure 1.1 National Forest System of the United States

Source: Department of Forest Sciences, Colorado State University, Fort Collins, Colorado.

Forest Management Policies in the National Forests

Much of the forest policy that developed in the United States during the twentieth century concerned the management and use of the National Forests. The U.S. Forest Service led the development of forestry research, education, financial assistance to state forestry organizations, and the expansion of responsible forestry practices on private lands.

From Custodial Management to Intensive Use

Before World War II, management of the National Forests was largely custodial. As noted previously, many of the eastern National Forests had already been heavily cut over and abused before the Forest Service acquired them. As National Forests they were protected from fire and further depredation during the decades it took for them to be restored through both natural processes and extensive replanting efforts. In the West, extensive wildfires had raged unchecked across thousands of square miles of National Forest in the summer of 1910, destroying millions of acres of valuable timber along with homes, ranches, and a significant number of lives. The 1910 fires shook the confidence of the young forestry agency and resulted in a committed effort to build an effective firefighting capability. This capability included not only training and equipping an army of wildland firefighters but developing new and innovative technologies for predicting, detecting, and suppressing wildfires. The image of Smokey Bear was first enlisted in 1944 for a massive public education campaign—one of the most successful in the nation's history—to carry the message of every citizen's personal responsibility for fire prevention. By 1952, Smokey Bear was on the cover of *Newsweek*, solidifying fire prevention as part of the core mission of the Forest Service. The policy of fire suppression throughout U.S. forests was so simple and direct, so successful, so well known, and so widely supported for so long that decades later, when ecologists had come to a clearer understanding of the important role of fire in many forest ecosystems, it was difficult for the Forest Service to persuade the public that forest fire could sometimes be a positive force.

In the 1940s, policies governing the National Forests shifted to a wartime stance and were aimed at helping supply the immense volume of wood needed to support the war effort. This new focus on wood production was accentuated during the postwar boom, when millions of wood-frame houses were being constructed in new suburban neighborhoods across the country. In earlier decades, the timber industry had sought to limit timber sales from the National Forests to keep from oversupplying the market and depressing timber prices. In the postwar years, however, with timber companies scrambling to keep up with the insatiable demand for lumber and plywood for housing, the industry clamored for the sale of an ever-larger volume of timber from the National Forests. Much of the forest industry land had been cut over and it would be decades before forests sufficient for harvesting would grow back. In short, the same industry that lobbied Congress to limit timber harvesting on National Forests prior to the war was now lobbying Congress to increase harvesting.

Congress readily complied with additional funding to support increased timber sales from the National Forests. Legislators saw more timber sales as a way not only to fulfill the need of an expanding nation for materials but also to create employment and income opportunities in forest industry–based communities in their home districts. Figure 1.2 shows the steady increase in timber harvesting in National Forests beginning shortly after World War II through 1988, a period of roughly 40 years. Congress was less receptive, however, to the other half of the Forest Service's message—that higher levels of timber harvesting had to be accompanied by greater investments in reforestation and other management activities aimed at maintaining healthy, growing forests. Congress regularly funded these activities at levels lower than the Forest Service requested, whereas timber sales were often funded at levels well beyond what was requested (Sample 1990). In many National Forests, this resulted in an increasing backlog of reforestation needs and led to conditions that in some instances appeared as bad as those the National Forests had been created to prevent (Hirt 1996).

Multiple-Use Forestry

A countervailing social trend in America during this period was the explosive growth in outdoor recreation, and in particular the recreational use of the National Forests. Americans with increased leisure time and the mobility provided by cars and im-

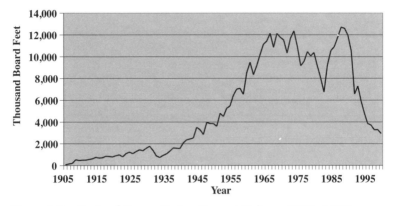

Figure 1.2   National Forest Timber Harvest Volume, 1905–1999

Sources: Timber harvest data from 1905 to 1960: M. Clawson. *The Federal Lands since 1956: Recent Trends in Use and Management.* Baltimore, MD: Johns Hopkins University Press for Resources for the Future, 1967, appendix table 3, p. 59. Timber harvest data from 1960 to 1980: M. Clawson. *The Federal Lands Revisited.* Baltimore, MD: Johns Hopkins University Press for Resources for the Future, 1983, table A-3, p. 283. Timber harvest data from 1980 to 1999: R. W. Gorte. *Timber Harvesting and Forest Fires.* Congressional Research Service Report to Congress. August 22, 2000. Available: http://www.cnie.org/nle/crsreports/forests/for-30.cfm, accessed April 6, 2003.

proved highways flocked to the National Forests in numbers that were regularly twice the numbers of National Parks visitors. Figure 1.3 shows how recreation visits for National Forests exceed those of National Parks over a 70-year time period, with the notable exception of the 1980s. Aesthetic and amenity resources of the National Forests took on far greater importance than ever before. The inevitable conflicts between recreationists seeking to experience wild Nature and loggers seeking the best and most accessible timber became more frequent and increasingly strident.

The Forest Service was eager to prove that it could manage such competing uses, however. The perception that the Forest Service focused only on timber production had been responsible for vast areas of the National Forest System being carved out to create National Parks—the Olympic and the North Cascades National Parks among them. Since 1916, when the National Park Service was established, the Forest Service had lost land to its counterpart agency and, as a result, had had to compete with it for budgets and prestige as an effective public land management agency. The Forest Service itself petitioned Congress to expand its mandate to include recreation, wildlife, and other nontimber resources. In 1960 the Multiple-Use Sustained-Yield Act became one of the most important forest policies governing the National

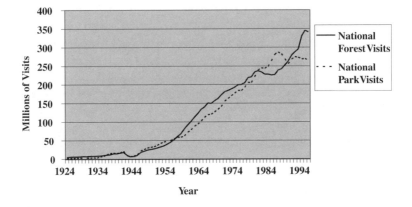

Figure 1.3   Recreation Visits for National Forests and National Parks, 1924–1996

Sources: Recreation visits for National Parks from 1924 to 1964: M. Clawson. *The Federal Lands since 1956: Recent Trends in Use and Management.* Baltimore, MD: Johns Hopkins University Press for Resources for the Future, 1967, appendix table 4, p. 60, and appendix table 36, p. 95. Recreation visits for National Forests from 1924 to 1996: Forest History Society. Available: http://www/lib.duke.edu/forest/usfscoll/policy/recreationvisitors.html, accessed April 6, 2003. Recreation visits for National Parks from 1965 to 1996: United States Department of the Interior National Park Service. Available: http://www2.nature.nps.gov/npstats/systemrpt.cfm, accessed April 6, 2003.

Forests since they were created in 1905. The Multiple-Use Sustained-Yield Act clarified that the National Forests were to be managed for wood production, livestock forage, water, fish and wildlife habitat, and recreation. Congress left it largely up to the discretion of the Forest Service to determine the appropriate balance among these various uses at any given time and location.

Wilderness

While reassuring to some, the discretion given to the Forest Service under the Multiple-Use Sustained-Yield Act further stimulated the growing public interest in the designation of wilderness areas to protect places of outstanding scenic value from the mechanistic incursions of modern life. Since 1924, the Forest Service had designated portions of the National Forests, often those that were remote and at high elevations, as wilderness areas. In these areas, it was the policy of the Forest Service to permit only primitive, nonmotorized kinds of recreation, and the focus of management was maintaining the wild character of the landscape by minimizing the evidence of human use.

These designations were made administratively by the Forest Service itself. With the ever-growing pressure on the Forest Service in the 1950s and 1960s to open new areas of the National Forests to logging, recreationists and wilderness enthusiasts became increasingly concerned that the Forest Service would reverse its earlier designations of wilderness areas and they would be lost. Aldo Leopold and Robert Marshall—both dedicated Forest Service employees—were among the leaders of a movement to have wilderness areas on National Forests and other Federal lands designated by Congress itself, taking the decision out of the hands of the agencies and making them more difficult to reverse. Leopold and Marshall were instrumental in creating The Wilderness Society, which today remains a strong voice for Wilderness designation on Federal public lands.

In 1964 Congress passed the Wilderness Act, designating more than nine million acres of Federal land as the National Wilderness Preservation System. Subsequent additions to the Wilderness system, particularly those contained in the Alaska National Interest Lands Act of 1980, have greatly increased its total area. Today, the National Wilderness Preservation System comprises more than 104.5 million acres. About 83 million acres in National Forests remain unroaded, and the question of which areas should be designated as Wilderness and which should be developed continues to be one of the most contentious issues in the management of these public lands.

The Clearcutting Controversy

By the 1970s, 30 years of intensive timber cutting, primarily through clearcutting (felling or removing every tree over an area of many acres), had taken a toll on the National Forests themselves and on the public's confidence in the Forest Service to conserve and protect these public resources. Public outcry over clearcutting was nothing new: Clearcutting in the Adirondacks by the Cornell School of Forestry in the late 1890s and early 1900s under the direction of Bernhard Fernow led to the dissolution of the school and to Fernow's emigration to Canada because of the public outcry against him. A series of citizen lawsuits against the Forest Service, notably the Izaak Walton League's 1974 suit against Monongahela National Forest in West Virginia, focused on the Forest Service's widespread use of clearcutting as the means of timber harvesting over large areas of land. The main thrust of these lawsuits was that by using clearcutting so extensively, the Forest Service had violated the Organic Administration Act of 1897. These citizen actions questioned the quality of the Forest Service's management of the National Forests and challenged the judgment of its forestry professionals. This produced a defensive reaction by the forestry profession in general, which drove an enduring wedge between foresters and the Conservation Movement that they had not only once been part of but for many years had led.

Congress responded in 1976 by passing the National Forest Management Act (NFMA). This legislation curtailed some of the discretion granted to the Forest Service by the Multiple-Use Sustained-Yield Act and prescribed specific forest management practices—particularly those relating to clearcutting—that would or would not be permitted on the National Forests. This new forest policy also established a complex process for developing ten-year management plans for each National Forest, requiring the Forest Service to involve the public in its decision-making and to adhere to all the public disclosure and environmental analysis requirements of the National Environmental Policy Act of 1970 (NEPA). NFMA proved once again that U.S. forest policy is an expression of social values as well as ecological considerations.

This planning process for each National Forest has become the venue in which competing interests press their often conflicting views about the appropriate balance of uses on these public lands. The regulatory requirements of other policies, although they are not strictly forest policies, must also be considered—NEPA, the Clean Water Act, the Clean Air Act, and the Endangered Species Act, to name a few. So complex are the analytical and legal requirements of this combination of process and regulatory policies that virtually any management decision can be held up for review by the courts and often blocked. This political gridlock in the management of the National Forests has given rise to many of the current issues in U.S. forest policy that will be discussed in chapter 2.

Establishing Sustainable Forest Management on Private Lands

Although much of the attention in forest policy continues to focus on the National Forests, these forests constitute only 19 percent of the nation's timberland area (Smith et al. 1997). Approximately

7 percent is in state, county, and tribal forests. The remaining 71 percent of the country's forests remain in private ownership— approximately 13 percent is owned by forest products companies and another 58 percent is in the hands of farmers and other private landowners. How to best accomplish sustainable forest management on the nearly three-quarters of the nation's forests that are privately owned has been one of the most persistent and urgent challenges in U.S. forest policy.

Evolution of the Forest Products Industry

The attitude of the early timber industry in the late nineteenth and early twentieth centuries toward sustainable forest management was driven by the harsh economics faced by any developing nation—being natural resource–rich but cash-poor, the nation exploited its resources to build wealth. Even while the Federal government was reserving millions of acres of public land because of concerns about a potential future timber famine, the timber industry still considered timber supply to be less a problem than other major factors of production—labor and capital. As long as labor and capital were scarce and expensive and trees were plentiful and cheap, it made little economic sense to the timber industry to invest labor and capital in reforestation or other aspects of forest culture. "Cut-and-run"—the practice of stripping the timber, abandoning the land, and moving on to other places where timber was still plentiful and land cheap—remained the unofficial forest policy of the lumber industry.

From the perspective of the forester and of communities left behind to cope with cut-over hillsides that could do little to hold back the spring floods, cut-and-run timber exploitation was no longer acceptable. With the timber industry unwilling to take the initiative to prohibit destructive methods of logging, a number of proposals were developed for making basic sustainable forest management practices mandatory on private lands, with either the Federal or state government in charge of enforcing these rules.

Like many other industries, the forest products industry in general has sought to minimize government regulation of any kind. However, some timber companies welcomed regulation as a way of stabilizing what had become a chaotic and unpredictable business environment. By the 1920s, some timber companies were already foreseeing timber supply problems and were shifting to a new approach—acquiring and holding land for the long term and investing in reforestation and other practices aimed at producing "second-growth" forests. There would be less waste in these forests because the companies would have greater control over species composition and could plant faster-growing trees of more uniform size. Companies among this group considered government requirements for basic, sound forestry practices—such as replanting after harvesting—to be largely consistent with their existing practices, which aimed to maintain land productivity to support successive crops of timber and to reduce the competitive advantage of less responsible timber operators. The key question was who should conduct such government regulation of forestry practices on private lands—the Forest Service or state forestry agencies. Lingering questions over the constitutionality of Federal regulation on private lands added to a general preponderance of public opinion in favor of state regulation. By the middle of the twentieth century, the question had for the most part been resolved in favor of regulation by the states (see chapter 2 for the history and development of state-level regulation of private forest practices).

The forest products industry has sought assistance through government policies aimed at improving the competitive advantage of U.S. companies internationally. The industry argues that such policies improve the economics of U.S. timber production, basing companies' adoption of better forest practices not on government regulation but on sound business decisions. In the view of the U.S. forest products industry, low prices for lumber and plywood (which limit profits and thus the industry's ability to invest in higher-cost forest management practices) are at least partly attributable to imports of cheap foreign lumber, chiefly from Canada. Historically, the influence of imports has been addressed through imposition by the U.S. government of import tariffs on wood products, especially softwood lumber. The first of these import tariffs was imposed in 1794. In March 2002, a 29 percent tariff was reimposed after negotiations to address the trade imbalance between the two nations broke down. Cross-border allegations of government subsidization and unfair trade practices continue to be a source of tension between Canada and the United States (see chapter 2).

Improving Management of Private Forests

State government regulation is only one of several policy mechanisms aimed at improving forest practices on private lands owned by individuals and organizations other than forest products companies. Other mechanisms include cost-sharing for activities such as fire protection and control of insects and disease, financial incentives for reforestation and silvicultural improvements (growing trees for wood production), and government purchase of conservation easements to reduce property taxes and protect private forests from development.

Forest taxation policy has long been at the heart of debates over how best to improve forest practices on private lands. Local property taxes and Federal and state "death taxes" have been blamed for making forestry uneconomical on small family holdings. Property taxes that increase each year with the appreciating value of both land and timber have been faulted with forcing premature or excessive timber harvesting, because landowners do not have the cash flow they need to pay the taxes until the timber is harvested. This problem has largely been addressed by local tax authorities shifting from ad valorem property taxation—taxing land according to its "highest and best use" fair market value—to "severance taxes" collected at the time mature timber is harvested.

Federal estate taxes, which are due within one year of an individual's death and can range as high as 50 percent, are blamed for forcing surviving family members to undertake excessive, unplanned timber harvests in order to pay the tax on forest assets. Others are forced to sell all or part of the land itself, contributing to fragmentation, development, or conversion to other nonforest land uses—all of which tend to degrade the public value of the forest land for wood production, wildlife habitat, watershed protection, recreation, and other uses. Federal estate tax reform has raised the amount of asset value that may be excluded from taxation to more than $1 million, but this exclusion is of little help to forest landowners with more than just a few acres. There are similar issues with state and local inheritance taxes, which are taxes paid by heirs on assets willed to them, rather than taxes paid by the estate of the deceased. Even under existing tax policy, individuals can significantly reduce these tax liabilities through comprehensive estate planning well before death. Assistance with estate planning is now one of the services offered to private forest landowners by private consulting foresters and state foresters.

Above and beyond the challenge of improving forest management on private lands is the challenge of conserving private forest land in the first place. Every year, hundreds of thousands of acres of private forest land are converted to nonforest uses such as development, effectively removing these lands from providing the multitude of goods, services, benefits, and values that intact forests supply. The effect has been twofold. First, private forests converted to nonforest land uses either lose their forest cover completely (such as when they are developed for residential housing) or have their tree cover diminished significantly. Forests that once protected water quality and prevented flooding by intercepting and storing rainfall often can no longer effectively serve this watershed-protecting function. Second, development in forested areas fragments the forest, both in terms of forest cover and of ownership. Increasingly, privately owned parcels of forested land are becoming smaller, raising challenges for integrated forest ecosystem management across a large area. It is far more efficient to manage one 1,000-acre parcel of private forest land for wildlife than it is to coordinate the management of 100 ten-acre parcels that are each owned by a different individual or organization. Private forest land fragmentation is a national trend from New England to the Rocky Mountain states, as urban dwellers flock to rural areas. Increased private forest land fragmentation and conversion to nonforest land uses pose some of the most daunting policy and management challenges, from wildlife and watershed conservation to managing wildland forest fires.

Conclusion

Forest policies in the United States have always been and continue to be expressions of social values. For much of the history of the United States, forest policies have been driven by fears of timber shortages, first at the local level and later at the national scale. Less than a century ago in the United States, wood was still the primary source of fuel for heating and cooking and was the raw material for many industrial processes; a sudden shortage would have significantly affected the national economy and indeed virtually every household. At the end of the nineteenth century, the United States made the transition from developing nation (with an economy based on resource extraction) to modern industrialized nation (with an economy based on agriculture, manufacturing, and international mercantilism), and forest policies were introduced as a way to ensure a reliable and perpetually sustainable supply of this important resource.

With the gradual shift to coal-based and then to petroleum-based energy sources by the mid-twentieth century, U.S. forests became less important as a source of fuel and industrial raw material. Forest policy then became focused almost entirely on ensuring timber supply for building products, paper, and wood fiber–based products. Enormous public investments were made to tame wildfires that burned millions of acres of forest each year, first on National Forests managed by the U.S. Forest Service, then by the forest products industry on its own timberlands, and finally on private forest lands through a variety of cooperative fire control programs operated by Federal and state forestry agencies.

The idea of sustained-yield forest management gradually replaced the practice of land clearing followed by abandonment. This shift was a matter of policy for National Forests and many state forests and was the objective of government efforts to regulate forest management practices on private lands. Achieving sustained-yield forestry created a demand for improved science and for technically trained foresters, resulting in substantial public support for forestry research and for forestry education at more than 50 colleges and universities throughout the United States.

During the final three decades of the twentieth century, public attention turned increasingly to the environmental values of forests, and forest policy development has reflected the strain that many see between the economic uses of forests and the protection of nonextractive forest ecosystem goods, services, and values. As debate around these issues has become increasingly adversarial and public consensus more difficult for policymakers to craft, the courts have played a greatly expanded role in determining how forests can or cannot be managed.

With this historical summary as background, the following chapters will describe some of the key aspects of forest policy in the United States today and provide a context for better understanding current issues and events in forest policy. This treatment is not intended to be either detailed or exhaustive. Instead, it is intended as a reference and a gateway to the much larger volume of information about forests and forest policy available in the printed scientific and professional literature; in electronic form on the Internet; and through the wide variety of public, private, and nonprofit organizations that are involved in forest conservation and the continuing evolution of forest policy.

References

Adams, D. 1993. *Renewable Resource Policy.* Washington, DC: Island Press.

Aplet, G., N. Johnson, J. Olson, and V. A. Sample. 1993. *Defining Sustainable Forestry.* Washington, DC: Island Press.

Dana, S. 1956. *Forest and Range Policy: Its Development in the United States.* New York: McGraw-Hill.

Hirt, P. 1996. *A Conspiracy of Optimism: Management of the National Forests Since World War Two.* Lincoln: University of Nebraska Press.

Lowenthal, D., ed. 1965. *Man and Nature: Or, Physical Geography as Modified by Human Action* by George Perkins Marsh (originally published in 1864). Cambridge, MA: Harvard University Press.

Pinchot, G. 1911. *The Fight for Conservation.* New York: Doubleday, Page and Company.

Pyne, S. J. 1982. *Fire in America: A Cultural History of Wildland and Rural Fire.* Seattle: University of Washington Press.

Sample, V. A. 1990. *The Impact of the Federal Budget Process on National Forest Planning.* Westport, CT: Greenwood Press.

Smith, W. B., J. S. Vissage, D. R. Darr, and R. M. Sheffield. 1997. *Forest Resources of the United States, 1997.* Gen. Tech. Rep. NC-219. St. Paul, MN: USDA, U.S. Forest Service, North Central Research Station.

USDA, U.S. Forest Service. 1993. *The Principal Laws Relating to Forest Service Activities.* Washington, DC: U.S. Government Printing Office.

Williams, M. 1989. *Americans and Their Forests: A Historical Geography.* Cambridge, England: Cambridge University Press.

Young, C. 1979. *The Royal Forests of Medieval England.* Philadelphia: University of Pennsylvania Press.

2

Current Issues in U.S. Forest Policy

Changes in forest policy in the United States today are issue-driven. The previous chapter described the historical development of U.S. forest policy over several centuries from European colonization of North America through the present day. Through U.S. history, there have been four distinct eras of forest policy—(1) nation-building and forest exploitation; (2) the rise of the Conservation Movement in response to growing fears of an impending timber famine; (3) development of scientific forestry and sustained-yield wood production; and (4) the environmental era and increased emphasis on ecological and amenity values of forests. During each of these periods, the basic framework of U.S. forest policy has reflected the prevailing social values and economic conditions in the nation as a whole.

Social values and economic conditions are not static. They continually evolve and change. With these changes in the societal context for forestry, and with improvements in our scientific understanding of forests, new issues arise and must be addressed through incremental adjustments in forest policy. If these incremental adjustments in forest policy are not made or are made too slowly, social pressure continues to build; some invisible political threshold is reached and seemingly overnight fundamental changes are made to the basic legal and institutional framework relating to forest conservation and management, ushering in the next era of forest policy.

This chapter provides a brief overview of some of the most

significant issues influencing current developments in U.S. forest policy:

- Sustainable forest management
- Conserving biological diversity
- Wilderness and roadless areas
- Forest health
- Community forestry
- Forestry regulation on private lands
- Forest land conservation
- Forest management certification
- Forests and water (watershed protection and water quality)
- Forests and atmospheric carbon
- Forest biotechnology
- Emerging issues and future policy directions

For each issue, this chapter provides essential facts, background, and a description of recent policy developments. This list of issues is by no means exhaustive or comprehensive, nor is it intended to address specific current legislative proposals or cases currently before the courts. Nevertheless, the issues discussed here reflect the range of social pressures currently building up in U.S. forest policy; they provide a context with which to better understand the daily ebb and flow of policy-related activities influencing the basic legal, institutional, and policy framework for the conservation and sustainable management of U.S. forests.

Sustainable Forest Management

Any study of forest policy must begin with the concept of sustainability in forest management. Operating under a variety of terms over more than two centuries—from "sustained yield" to "sustainable forestry" to "forest ecosystem sustainability"—the central concern of forestry has always been essentially the same: how to meet current needs for wood, wildlife, forage, and other products of the forest without diminishing the forest's basic productivity and its ability to continue meeting those needs in the future.

Sustainability in forest management is both a biological and socioeconomic concept. Originating in eighteenth-century Europe with the aim of avoiding social and economic disruptions associated with local timber shortages, sustained-yield forest management evolved to a highly technical process of modeling growth, mortality, and risk in order to set timber removals at a level that theoretically could be maintained in perpetuity (annual forest removals should equal annual forest growth minus annual mortality). Changing scientific understanding of the ecological functioning of forest ecosystems has challenged the notion that a sustained yield of timber is equivalent to sustaining all the components and natural processes necessary to maintain the long-term health and productivity of these ecosystems.

There is widespread public debate currently, both in the United States and elsewhere, as to what constitutes sustainable forestry. The continuously evolving science of forest ecology, as well as the continuously evolving social values and perceptions relating to forests, mean that what constitutes sustainable forest management cannot be defined once and for all time. The concept of sustainability itself will continue to evolve, and so must the laws and policies designed to promote sustainable forest management. Ensuring that forest policy is flexible and dynamic enough to accommodate evolving science and changing societal expectations is critically important for guiding the development of future forest policies that will be both durable and effective.

Background

Sustainability has always been a primary focus of forestry. Indeed, it was concern that forest resources would be inadvertently depleted, leading to unacceptable social and economic effects, that first gave rise to the systematic study of forests and a scientific approach to the long-term management of the forests' resources.

Sustainability in forest management began as both a biological and socioeconomic concept. Foresters developed an understanding of natural forestry productivity—and how it might be enhanced through silviculture—to maintain a continuous supply of wood, game, and other products for human use and consumption. The concept was fundamentally driven by the desire to avoid the social and economic disruption associated with shortages of timber, whether for local use or as the basis for a community export economy. Forest products clearly held the potential of being a perpetually renewable resource, and foresters undertook the responsibility of making this so.

As mentioned previously, the origins of sustained-yield forest management can be found in eighteenth-century Europe (Heske 1938). The lack of well-developed systems for transportation and communication were barriers to the development of regional trade (Waggener 1977). Local consumption was almost entirely dependent on local production, and communities had to be largely self-sufficient. There was a distinct possibility of exhausting local timber resources unless collective use was strictly controlled, and so the production and consumption of forest products became communally regulated. Perhaps because of the opportunities it afforded for employment and income in rural communities, this approach to sustained-yield forestry persisted long after improved transportation and communication systems had reduced the need for local self-sufficiency and turned timber into an ordinary economic commodity.

It was in this context that the concept of the *regulated forest* came into being. Under this concept, forests were to be managed to yield periodic, regular, and sustainable timber harvest volumes. In theory, the objective of maximizing the volume of timber that can be sustainably harvested from the forest is achieved by harvesting when the annual increment of growth in the stand reaches a maximum ("culmination of mean annual increment"). This approach recognizes that trees begin their growth slowly, with only modest increases in annual volume; then increase their growth at a faster rate; and finally, beyond some age, begin to experience a decline in annual additions to tree volume. The "culmination of mean annual increment" rule gives the rotation age at which the sustainable harvest volume would be maximized. The harvest level is determined using "Hanzlik's Formula," which divides the net growth over the entire area of the economic enterprise by the rotation length and indicates the average annual volume of timber that can be removed on a sustainable basis.

In the mid-nineteenth century, it was shown that forestry enterprises seeking to maximize financial returns would, in most cases, harvest younger trees on shorter rotations (Faustmann 1849). Although harvesting at the culmination of mean annual increment maximized the physical volume of the harvest, it could result in lower financial returns from the forest, because it did not account for the cost of capital tied up in land and forest management expenses. This concept was controversial at the time and remains so today. In Europe, where centuries-old forest enterprises have endured through wars, currency devaluations, and other events that have put other forms of capital investment at great risk, forests have served as a stable, reliable, tangible asset. Increasing long-term financial security can be as much a concern as near-term financial returns. Building value in relatively low-yielding forest assets (compared with investments in stocks or bonds) reduces risk in a diversified portfolio of investments and thus increases overall return (Binkley 1996).

Recent Policy Developments

Sustained-Yield Forestry in the United States

These concepts of sustained-yield forest management were transplanted to the United States at a time of growing concern over the possibility of a timber famine—nationally as well as locally. Forests in the United States had been regarded as both an inexhaustible resource and an obstacle to the westward expansion of agriculture. By the late nineteenth century, wood was still the major building material and the predominant source of fuel. Vast areas of forest had been cleared but not reforested, and there was a very real concern that a timber shortage would begin to seriously limit the prospects for future economic growth.

As introduced in the United States by early foresters Bernhard Fernow, Gifford Pinchot, and others, forestry was largely a technical undertaking. It was broadly assumed that by maintaining a continuous supply of timber and protecting the basic productivity of soils and watersheds, a broader set of forest uses and values would automatically be protected for the American people as a whole. Federal forest reserves were established by the Forest Reserve Act in 1891. The purpose of the Federal Forest Reserves, later called the National Forest System, was to provide "the greatest good, for the greatest number, in the long run" (Pinchot 1947).

Custodial Management

Management of the National Forests was largely custodial until the mid-1940s. Preventing theft and wildfire was the major activity in the National Forests of the western United States. In the East, large areas of cut-over forest and abandoned farmland were acquired by the U.S. Forest Service and gradually restored through replanting, erosion control, and land stabilization efforts. Conversion of forest to other land uses was generally prohibited. Little timber was cut on the National Forests during this period, partly because political pressure from timber companies sought to minimize competition in the private wood products industry and maintain favorable prices for private timber. Management of public forests emphasized maintaining the land in its native forest cover and relying on natural regeneration following disturbances. The underlying biological and ecological systems were not well understood, however, as evidenced by the way wildfire was viewed at the time. Rather than recognizing that wildfire was part of a natural disturbance regime integral to the functioning of the forest ecosystem, policymakers of the day sought to suppress wildfires whenever and wherever they occurred. Thus, even custodial management requires a thorough understanding of natural disturbance regimes and other complexities of forest ecosystem functions.

By the 1940s, many private timber companies had also come to accept the idea of sustained-yield forest management. Previously, the standard practice had been to acquire forest land, liquidate the timber assets, and abandon the land—an approach known as "cut out and get out" or, more succinctly, "cut and run." Under the leadership of corporate pioneers such as Frederick Weyerhaeuser, private timber companies began to recognize the benefit of holding land, reforesting it, and harvesting timber on a renewable basis. Today, the management of many private forest lands in the U.S. reflects the sustained-yield forestry of nineteenth-century Europe, although using modern technology and adhering to certain environmental constraints. With private timber supplies drawn down by the war effort, the National Forests in the late 1940s and 1950s became a major supplier of timber for economic expansion and the suburban housing boom. Increased leisure time and improved transportation systems brought more Americans in contact with the National Forests, increasing demand for recreation, wildlife, and other noncommodity resource values. With growing frequency, large-scale timber harvesting activities came into conflict with these other uses, challenging the operational utility of the traditional concept of sustained yield as the maximization of timber yield constrained only by the biophysical limits of the land itself.

Multiple-Use Forestry

The Multiple-Use Sustained-Yield Act (1960) was an important turning point in foresters' interpretation of their responsibility for sustainable forest management. It defined sustained yield as "the achievement and maintenance in perpetuity of a high level annual or regular periodic output of the *various renewable resources* of the National Forests without impairment of the productivity of the land" (16 USC 528 1960, emphasis added). It has long been recognized that forests generate a host of goods and services simultaneously. Medieval forests were commonly valued for their game and forest foods, as well as for wood used as fuel and construction materials (Westoby 1989). Even when forests are managed for timber, other values are commonly produced as by-products. Fish and wildlife, recreation, water and water quality, range, and other outputs are commonly generated incidentally to the production of timber. It was this notion of producing multiple benefits for multiple uses that inspired the Multiple-Use Sustained-Yield Act.

The Multiple-Use Sustained-Yield Act provided the statutory basis for the application of this approach to U.S. public forests. Public controversies over the Forest Service's implementation of multiple-use forestry have led to additional statutory direction for sustainable management of the National Forests. The Forest and Rangelands Renewable Resources Planning Act of 1974 (RPA) required periodic national assessments of the supply and demand for a large array of resource uses and values—and a strategic plan detailing how the Forest Service intended to address all demands simultaneously (16 USC 1600). The agency's answer, in a word, was *money*; with significant increases in funding, investments in intensive resource management would allow all the renewable resources of the National Forests—several of which, like timber production and providing certain kinds of wildlife habitat, or timber production and certain forms of recreation, competed and conflicted with one another—to be sustained indefinitely (Sample 1990).

In the decades following the passage of the Multiple-Use Sustained-Yield Act, the public grew increasingly dissatisfied with the balance the Forest Service had struck in addressing these competing sustainability goals. The predominant focus on timber production that had developed in the agency during the 1950s persisted, largely because budget allocations from Congress favored sustaining high harvest levels over other forest management goals. Public criticism suggested that such high levels of timber removal not only imposed unacceptable effects on the nontimber resources but threatened the long-term sustainability of timber production as well. The Forest Service's optimism and estimates of the increasing level of sustainable timber harvesting were based on technical assumptions that overlooked an important fact—that needed investments in intensive forest management were simply not being funded by Congress and thus were not being made (Hirt 1994).

Nondeclining Even Flow

In 1976, the National Forest Management Act (NFMA), an amendment to RPA, placed numerous additional statutory limits on timber production on the National Forests, and it required the development of detailed management plans with ample opportunity for public involvement in decision-making (16 USC 1600 [note]). Many of these limitations were aimed at reducing the effects of timber harvesting on nontimber resources such as wildlife or recreation. But concern over the sustainability of timber production itself led Congress to add a new wrinkle to its definition of sustained yield: it specified that the sale of timber from each National Forest be limited to "a quantity equal to or less than a quantity which can be removed from such forest annually in perpetuity" (16 USC 1600, 13). This so-called "nondeclining even flow" constraint was criticized by some economists as inherently inefficient in managing the extensive areas of native forest old growth that existed on many western National Forests at the time (Clawson 1983). Old-growth forests, especially in the Pacific Northwest, can reach the age of 800 to 1,000 years and, therefore, far surpass the point of culmination of mean annual increment (the age at which harvest is typically set for maximizing timber production). Because of the high rate of tree mortality, old-growth forests can actually lose timber volume faster than they put it on. Purely from the perspective of sustained-yield forest management, old-growth forests are very inefficient and need to be converted to young vigorous and fast growing forests. The debate over nondeclining even flow was effectively ended in the early 1990s when there was a virtual moratorium on harvesting old growth to protect habitat for the northern spotted owl and other species dependent on old-growth forests.

Taken together, the Multiple-Use Sustained-Yield Act and the National Forest Management Act represent an important shift in the concept of sustainability in forest management—from the maximization of a single objective subject to ecological and environmental constraints, to the simultaneous pursuit of multiple objectives. Whereas sustained-yield forestry has as its sole objective the maximization of timber volume production (or the maximum financial returns), the intent of multiple-use forestry is to satisfy numerous resource output objectives simultaneously. In developing plans to implement these laws for National Forests, the Forest Service took a highly technical approach, using complex mathematical models to determine which mix of products and "nonmarket values" could be expected to maximize "net public benefits." In practice, this mechanistic approach simply did not work, as evidenced by the political and legal gridlock that came to characterize National Forest policy and management. Additional concerns over endangered species led to sudden and immediate court-imposed reductions in timber supply in some areas, resulting in a political impasse and a fundamental reexamination of what forest managers are to sustain, for whom, and to what purpose.

Definition of Sustainable Forest Management

This reexamination is leading to a further evolution in the definition of sustainability in forest management, one that explicitly rather than implicitly includes social and economic, as well as biological, objectives. A key tenet of sustainable forest management is that it must be not only ecologically sound but also economically viable and socially responsible (Aplet et al. 1993). If it is lacking in any one of these three areas, the forest management approach will not be tenable in the long run.

Most conservation interests now acknowledge that it is impossible to accomplish long-term protection of forest ecosystems without incorporating into conservation strategies the economic and social needs of the local communities. Economic development and commercial interests are recognizing that ensuring the ecological soundness of their activities helps not only to ensure raw material supplies for the future but to maintain essential social and political support (Schmidheiny 1992). Communities are no longer willing to accept the social disruptions and family dislocations that have always accompanied a boom-and-bust approach. They are recognizing that government policymakers alone cannot lead the way toward stable, resilient, and economically diverse communities—that there is an important role for partnerships among Federal and state governments, business interests, and the communities themselves in finding a new basis for sustainable resource use and sustainable communities.

The debate over sustainable forest management in U.S. forest policy is not taking place in isolation; it is part of a larger international effort to define, articulate, and encourage the practice of sustainable forestry. At the 1992 United Nations Conference on Environment and Development (UNCED) in Rio de Janeiro, consensus was reached on a set of nonbinding "forest principles" defining sustainable forest management. Since that time, the United States has made significant progress toward a nationwide assessment of forest conditions and trends relative to an internationally agreed-on set of criteria and indicators of sustainable forest management developed at a 1993 conference in Montreal (the Montreal Process). Both public and private forest interests in the United States are currently striving to implement a series of proposals for action aimed at improving forestry, as measured against the criteria and indicators of sustainable forest management. For example, the American Forest and Paper Association has a Sustainable Forestry Initiative (SFI) based on a derivation of the Montreal criteria and indicators of sustainable forest management. Similarly, the Forest Service and several state forestry agencies are working on their own set of criteria and indicators for sustainable forest management in their respective jurisdictions.

Conclusion

Sustainability in forest management is a dynamic, evolving concept, reflecting changing social values and the evolution in our scientific understanding of the effects of human activities on the functioning of forest ecosystems. As an increasingly broad cross section of forestry interests comes to accept that truly sustainable forestry must reflect ecological, economic, and social objectives, the most challenging trade-off for policymakers may be between short-term needs and long-term assurances.

The central idea behind "sustainable development"—that is, meeting the needs of current human society without unduly compromising the capacity of future human societies to meet their needs (World Commission on Environment and Development 1987)—is not materially different from the basic motivating concept behind sustained-yield forestry in eighteenth-century Europe or sustainable forestry in twenty-first-century America.

From a policymaking and operational management perspective, it seems the sustainability challenge will always be to protect the long-term productivity of forest ecosystems—to the best of our biological, social, and economic understanding—without unduly limiting the use of forests to meet current needs. Or from an analytical perspective, the challenge will be to operate as closely as is socially and politically acceptable to the limits of what can be sustainably produced, neither exceeding ecological capacities nor leaving significant ecological capacity unused. How conservative a margin for error is incorporated is as much a political decision as a scientific one.

Conserving Biological Diversity

The conservation of biological diversity, and especially the protection of habitat for threatened or endangered species, has emerged as one of the foremost issues in forest management both in the United States and around the world. Although forest managers have been able to adapt to the need to provide an ever-broader array of forest values and uses, the increasing need to conserve biological diversity may be the most challenging yet. The downward trend in biodiversity and the potential of forest protection to slow that decline are seen by many as sufficient reason to cease any and all forest management activities that potentially interfere with that objective.

Background

Many of the world's most recognized and respected biologists believe that we are now in the midst of a biodiversity crisis, with extinctions of animal and plant species taking place at a rate not seen since the dinosaurs were wiped out 65 million years ago. Harvard biologist Edward O. Wilson has estimated the current rate of species extinctions at approximately 27,000 per year—or an average of 74 each day—out of a worldwide total of perhaps 10 million species (Wilson 1992). The normal "background" extinction rate is about one species per one million species a year (Raup and Sepkoski 1984). More than 20,000 species are globally rare or threatened, and as many as 60,000 face extinction by the middle of this century (IUCN 1988). According to Wilson, "Human activity has increased extinction between 1,000 and 10,000 times over this level . . . clearly we are in the midst of one of the great extinction spasms in geological history" (Wilson 1992, 280).

The world's greatest concentration of biological diversity in forest ecosystems—and the greatest threats to conserving that diversity—are in the tropics (Raven 1987). Because of the means by which tropical rainforests cycle their nutrients, these seemingly lush and irrepressible forests are much more vulnerable to ecological damage than most temperate-zone forests and much slower to recover from deforestation (Wilson 1992, 274). The growing loss of forest area in the tropics is the single greatest threat to global biodiversity, a trend that is exacerbated by population growth rates in many tropical nations that far exceed those in most temperate-zone nations. "An awful symmetry binds the rise of humanity to the fall of biodiversity: the richest nations preside over the smallest and least interesting biotas, while the poorest nations, burdened by exploding populations and little scientific knowledge, are stewards of the largest" (Wilson 1992, 272).

The importance of conserving biological diversity in forest ecosystems has generated policy proposals aimed at minimizing the conversion and fragmentation of the remaining large areas of native forests and preventing the diminishment of remaining biological diversity by development for commodity production. Wilson estimates that the 4.3 percent of world's land surface currently under legal protection should be expanded to 10 percent (Wilson 1992, 337). Many eminent biologists and other scientists support a proposal to set aside 50 percent of the North American continent as "wild land" for the preservation of biological diversity (Ehrlich 1997). Environmental organizations in the United States, such as the Sierra Club, are actively working to ban all commercial timber harvesting on Federal public lands.

Many conservation biologists today point to the need to think beyond the "reserve mentality" in designing strategies for conserving biological diversity (Brussard, Murphy, and Noss 1992). But it is also clear that forest reserves will continue to be a major component of any successful biodiversity conservation strategy (Hunter and Calhoun 1996), particularly with regard to species endemic to late-successional forest ecosystems—that is, old-growth forests (Spies and Franklin 1996).

The global nature of the biodiversity crisis points up the need for a strategy that integrates the management of temperate, tropical, and boreal (northern) forests with world demand for wood. Current global industrial roundwood demand is estimated at 1.6 billion cubic meters per year and is expected to rise to 2.5 billion cubic meters per year by 2050 (FAO 2000). Industrialized nations account for a disproportionate share of this global demand, and among the developed nations, the United States stands out as one of the world's largest consumers of wood. U.S. per capita consumption of major wood products (lumber, plywood, and paper) is about double that of Germany, 7 times that of Brazil, and 15 times that of China (FAO 2000). The United States has one of the lowest average population densities among the developed nations (for example, Oregon has a population of fewer than 3 million people; Germany, with a geographic area slightly larger than that of Oregon, has a population of more than 82 million) and some of the most productive forests. In spite of this, the United

States continues to import more than one-quarter of its wood— 114 million cubic meters in 1997—from harvesting in both tropical and boreal forests outside of the United States (Howard 1999).

Although temperate forests are comparatively less biologically diverse, they have "hot spots" with extraordinary concentrations of species diversity, particularly where there are large, contiguous areas of mostly undisturbed native forest (Ricketts et al. 2000). For a wealthy, temperate-forest nation like the United States to support a credible and ethical program for biodiversity conservation in the poorer tropical nations, its own policies for sustainable forest management are gradually moving toward a two-pronged strategy of (1) protecting their own biodiversity hot spots where they exist, even when it means sacrificing economic values that could have been derived through resource development, and (2) sustainably using productive forest areas of relatively low biodiversity value to help alleviate the pressure on tropical and boreal forests to meet global needs for wood fiber and other renewable resources.

Recent Policy Developments

The single most important policy with regard to conserving biological diversity is the Endangered Species Act of 1973 (ESA). Under the terms of ESA, the United States is acting to protect habitat for threatened and endangered species on both public and private forest lands and to prohibit the trade of threatened and endangered species internationally. Controversy abounds on both public and private lands, however, and it will be increasingly difficult to guarantee the survival of every species without limiting economic activities in forests in ways that many would find difficult to accept.

Public Lands

The Endangered Species Act requires the managers of Federal forest lands to protect habitat for species that have been listed by the U.S. Fish and Wildlife Service (USFWS) as threatened or endangered. When a species has been listed as threatened or endangered, critical habitat is often designated. Critical habitat is defined as geographic areas with features that are essential to the survival of a species and that require special management considerations. Regardless of critical habitat designation, the agencies develop a recovery plan that serves as a coordinated strategy for increasing the species' population to the point where it is no longer considered threatened or endangered. Federal land management agencies such as the Forest Service or Bureau of Land Management are required to consult with biologists from the wildlife agencies before they undertake any management activities that could have potentially negative effects on the habitat of listed species.

Although it is quite logical in principle, the required consultation process can become a substantial source of delay and additional project planning costs for the land management agencies. In recent years, the wildlife agencies have been so overwhelmed by the number of projects on which consultation is needed that it sometimes takes years for a proposed project to be reviewed and approved. Often interest groups then take legal action to appeal the decision, whichever way it is rendered, which means it may be many more months or even years before a final decision is made on whether or not the project may proceed. Several policy proposals to streamline the consultation process have been considered, but they are often opposed by citizens' groups because they are regarded as limiting public review of agency decisions or increasing the likelihood that potentially damaging aspects of a proposed project will be overlooked in haste.

In the case of the National Forests, the habitat conservation requirements of the National Forest Management Act have been interpreted as far more stringent than those of the Endangered Species Act. The administrative regulations implementing NFMA have required that the Forest Service ensure viable populations of species throughout their range (36 CFR 219). This requirement is more rigorous than those of ESA in two ways. First, NFMA regulations require that the Forest Service ensure the viability of the species themselves, whereas ESA requires only that suitable habitat be maintained. Second, the NFMA regulations require that these populations be maintained throughout their range, whereas it would be permissible under ESA for a regional subpopulation to decline as long as adequate habitat is maintained overall under a recovery plan.

The fact that the Forest Service can directly influence the management of only the National Forest lands—and not adjacent lands that also include habitat for listed species—has been used to support the view that ensuring the viability of the species itself is an impossible standard to which the Forest Service should not be held. This argument has been made, not least of all, by a former Forest Service Chief who is among the nation's most respected conservation biologists. In recent years, there have been several proposals to revise the NFMA regulations to, among other things, bring them into uniformity with the requirements of the Endangered Species Act in regard to maintaining suitable habitat for threatened or endangered species. But it has been argued, and quite correctly, that this would constitute a significant reduction in the protection currently given to listed species in National Forests. How this will be resolved is much less a scientific or technical issue than it is a political issue: Whose interests will be helped, whose will be hurt, and who can prevail in the policy arena?

Private Lands

Biodiversity conservation on private forest lands is handled quite differently under ESA. The act prohibits the taking of any listed species—defined not only in terms of directly harming the animal, but of indirectly harming it through destruction of critical habitat. Essentially, if critical habitat for a listed species occurs on private forest land, that habitat cannot be significantly modified through any means, including timber harvesting. To many private forest landowners, this represents a "taking" of private property without due compensation, which is prohibited under the Fifth Amendment to the U.S. Constitution. In some instances, the Federal courts have agreed. In most cases, however, courts have found that protecting the habitat in question does not deny a significant enough portion of the owner's economic use of the land to constitute a taking (Stedfast 1999).

Nevertheless, several options are available to private forest landowners that, in many cases, will allow them to continue actively managing their forest land for their own objectives. Landowners can work with USFWS biologists to develop habitat conservation plans designed to allow timber management and other activities to proceed while still conserving habitat for the resident population of the listed species present. Adhering to the terms of the plan assures the landowner that, should any subsequent activities result in an "incidental take" (the accidental death of an individual of this species), the landowner is not liable. Experience has shown the development of a habitat conservation plan to be a costly and time-consuming process that, although a useful option for a forest products company or large private landowner, is not financially feasible for the majority of small private forest landowners.

Another option developed specifically with small landowners in mind is the safe harbor agreement. With the increasing attention being given to threatened or endangered species with a preference for late-successional or old-growth forest habitat, private landowners have been known to harvest their timber prematurely rather than take the risk that a listed species will be found to have taken up residence in their forest, thereby limiting their management options. Under a safe harbor agreement, the population of the listed species is determined at the time the agreement is reached, and this becomes the population level the landowner is subsequently responsible for maintaining—no matter how much the population might increase thereafter. This effectively removes the risk to the landowner of allowing the forest to continue to maturity, providing a higher financial return to the landowner when the harvesting takes place and providing critical habitat to late-successional forest dependent species in the meantime.

Forest Plantations and Protected Areas

Conserving biodiversity in managed forests limits the level of timber management activity and the degree of ecological intervention it represents in forest areas containing critical habitat. For numerous species, particularly those found in interior regions of large late-successional forests, even minimal levels of ecological disturbance result in significant diminishment of habitat quality and perhaps loss of the species. On the other hand, there are areas of forest land that are of relatively low value for biodiversity conservation but are highly productive and well-suited in other ways to sustainable wood production. There is a distinct possibility that taking a more specialized approach to managing areas especially well-suited to either biodiversity conservation or sustainable wood production will result in a net gain in both wood production and the protection of threatened and endangered species.

It is now widely recognized that for many uses, wood is environmentally preferable to other building materials such as aluminum, steel, or concrete. All of these other materials require far more energy to convert the raw material into usable product, and far more waste is generated, with important implications for the consumption of fossil fuels and for emissions of carbon dioxide and other greenhouse gases. These nonrenewable resources leave degraded landscapes, mine wastes, and other chemical pollution related to processing and manufacturing. By comparison, wood from well-managed forests is perpetually renewable, easily recyclable, and ultimately biodegradable. Energy requirements for processing wood are relatively low and often are reduced further by cogeneration using non–fossil fuels (such as sawdust and chips) from the wood and paper manufacturing processes themselves.

A recent report by the World Wildlife Fund (WWF) suggests that a significant expansion of the area of intensively managed forest plantations could allow the world's major forest products companies to meet a substantial share of the global demand for industrial roundwood from a relatively small proportion of the world's forest area. The report also suggests that such an expansion of intensively managed forest plantations could open up new opportunities to provide outright protection to forests with high conservation values, particularly those with globally significant biodiversity (Howard and Stead 2001). WWF is so convinced of the value of this approach that it has called on the world's ten largest forest products companies to collectively increase the area of intensively managed forest plantations by 5 million hectares per year for the next 50 years (World Wildlife Fund 2001). With this level of investment, WWF estimates that as much as 80 percent of the world demand for industrial roundwood in 2050 can be met from less than 20 percent of the world's forests. Furthermore, WWF asserts this can all be done in a manner that is consistent with the Forest Stewardship Council (FSC) criteria for green certification—meaning that much of that 20 percent will be new forests planted on retired marginal crop and pasture land rather than plantations created by converting natural forests.

What would such an approach mean for the United States? In some ways, we are already tending in this direction: Wood production is shifting increasingly to private industrial timberlands and other private forests, and biodiversity conservation is becoming a primary management goal in many public forests. Nevertheless, significant policy and political barriers remain to achieving either of these objectives efficiently or effectively. We are perhaps within reach of a new political consensus—one in which both the forestry community and the environmental community actively support the idea that intensively managed forest plantations *and* protected areas in high-conservation-value forests have an essential place in a comprehensive strategy for sustainable forest management. New Zealand has already gone this route, and other countries are seriously considering this approach in the face of competing demands on their forests.

Policymakers have an opportunity to further develop this

potential for broad public consensus on forests and forestry and to shape a policy framework that will support and facilitate this kind of practical approach to accomplishing sustainable forestry. It has been suggested (Binkley 2001) that a consensus agreement might include considerations such as these:

- Devoting 20 to 30 percent of the land area of plantation projects to ecological services.
- Maintaining strict control of offsite impacts of plantation-based timber production, especially the movement of silt, fertilizer, or herbicides into waterways or groundwater.
- Establishing agreed-on limits regarding the use of yield-enhancing chemicals such as fertilizers and herbicides, focused on minimizing their use and maximizing their effect.
- Establishing agreed-on limits regarding the use of genetically modified organisms to instances in which it can be demonstrated that gene flow out of the plantation is impossible.
- Making a commitment not to log old-growth forests.

Numerous additional limited opportunities exist to create a policy framework that enables and encourages public and private forest land managers to make rational choices that will tend to be consistent with and supportive of this general approach. For example, a more expansive approach to the Conservation Reserve Program (a Federal assistance program for private landowners) could provide landowners with the incentive to turn marginal crop and pasture land into productive forest plantations. Raising or eliminating the $10,000-per-year cap on the Reforestation Tax Credit would give new income tax incentive for undertaking the significant up-front costs of establishing forest plantations. Expanded use of safe harbor agreements under the Endangered Species Act could insulate landowners who establish forest plantations through afforestation, providing greater assurance that they will be able to recoup their investment.

Conclusion

The world is losing species at a faster rate now than at any time in human history. The conservation and sustainable management of forests, both in the United States and in other regions of the world, holds one of the most important keys to slowing this loss of biological diversity. As one of the world's most heavily forested nations—and as one of the world's largest consumers of wood products—the United States has a dual conservation responsibility to fulfill: The United States could protect its remaining hot spots of biological diversity—and bear its share of the local, short-term economic effects of doing so—and at the same time meet its share of the demand for renewable wood and fiber that the United States itself generates without shifting an undue burden on biologically rich forests in other regions of the world. Joining together the interests of the environmental community, the forest industry, and government could become the basis for the broadest and most effective consensus on U.S. forest policy in half a century. Policymakers are just beginning to consider the mechanisms that will facilitate this process, but it is one that is certain to receive significant attention in coming years.

Wilderness and Roadless Areas

The idea of preserving large areas of wilderness for their own sake sprang from the uniquely American experience. European colonizers arrived on what they believed was an untamed continent and struggled over a period of more than two centuries to bend the landscape to their will and make it look like the ones their ancestors had left behind. It was not until the late nineteenth century, when most of the continent had fallen under the ax or the plow, that Americans began to recognize the important role that wilderness had played in shaping American culture, values, and perspectives. Wild, untamed landscapes became valuable in their own right—not because they looked like European landscapes, but precisely because they did not. North America was one of the few places in the known world at the time that could boast such spectacular places as Yosemite or Yellowstone. Americans took a new pride in their wild landscapes and felt a new fascination with protecting these pieces of their national heritage.

Background

Yosemite Valley was transferred to the ownership of the state of California in 1864 and was designated a state park in 1869. It later became Yosemite National Park in 1890, but the core valley was not transferred to the Federal government until 1906, when it was officially added to the National Park System. Yellowstone was withdrawn from development in 1872 and was protected by a contingent of the U.S. Army. Both became "crown jewels" in the National Park System when it was formed by Stephen Mather in 1916, becoming a model for national parks throughout the world. Many of these parks, managed as "pleasuring grounds for the public," came to feature their own kind of development, complete with roads, parking lots, hotels, and restaurants. In the 1930s, conservationists began looking to other areas of undeveloped Federal lands, primarily the western National Forests, to protect landscapes "untrammeled by Man . . . where Man himself is a visitor who does not remain" (16 USC 1131 [note]).

Some of these areas were protected by the Forest Service itself. With the help of Forest Service leaders like Aldo Leopold and Robert Marshall, the agency designated Wilderness areas as early as 1924 for their backcountry recreation opportunities and spectacular scenery and as a baseline for research on the ecological effects of human activities on other more-developed landscapes. The first roadless area review by the Forest Service was in 1929 (under the "L-20" regulations); a second review (under the "U-Regulations") took place in 1939. With each review, the size of the areas under review became smaller. With the new emphasis on timber production in the National Forests in the years following World War II, conservationists became increasingly concerned that these administratively designated Wilderness areas could just as easily be "de-designated" by the Forest Service and made available for logging. Logging, in addition to its immediate effects on the landscape, requires the construction of an extensive road network that, long after the logging is completed, still allows other human activities to permeate the area—motorized recreation vehicles, increased hunting pressure, and human-caused wildfires. Conservationists felt that an act of Congress would be required to truly protect the remaining areas of wilderness on Federal lands.

The Wilderness Act of 1964 immediately protected more than 9 million acres of wild areas on lands managed by the Forest Service, National Park Service, U.S. Fish and Wildlife Service, and the Bureau of Land Management, primarily in the western continental United States. Subsequent congressional acts designated additional Wilderness in the eastern United States (1978) and Alaska (1980). In 1971 the Forest Service undertook a Roadless

Area Review and Evaluation (RARE) to evaluate the wilderness potential of every unroaded area in the National Forest system larger than 5,000 acres—a total area of almost 56 million acres. Concerned that the Forest Service would quickly develop some of these areas to disqualify them for future Wilderness designation, the Sierra Club filed a lawsuit *(Sierra Club v. Butz)* that resulted in a requirement that the Forest Service prepare an environmental impact statement detailing the values that would be lost through development. A second roadless area review (RARE II) was initiated, but in the meantime the passage of the Eastern Wilderness Act essentially broadened the operational definition of *Wilderness* by designating areas as small as the 2,570-acre Gee Creek Wilderness in Tennessee.

The RARE II inventory was thus expanded to include 62 million acres of National Forest. In the end, the Forest Service recommended only 25 percent of these lands (15.4 million acres) for future Wilderness designation; 17 percent (10.6 million acres) were recommended for further planning, and the remaining 58 percent (36 million acres) were to be reopened for timber production, grazing, mineral development, and other uses. A lawsuit challenging the adequacy of the environmental impact statement prepared by the Forest Service temporarily halted new development on any of the inventoried roadless areas, however. Even today, the fate of large roadless areas in places such as Montana and Idaho is the topic of contentious debate.

Since the RARE II inventory, Congress has designated new Wilderness areas on a state-by-state basis, and with each bill development interests sought the permanent release to multiple-use management of all roadless areas not being designated in the bill as Wilderness. Many of these Wilderness areas received their designation only after traditional resource uses such as grazing and mining were grandfathered in as acceptable. Hence, even today, grazing and mining can occur in designated Wilderness— the price of political compromise. In all, an additional 4.4 million acres were designated as Wilderness in the continental United States, and none of the remaining roadless areas were legislatively released. The Alaska National Interest Lands Act in 1980 added another 5.3 million acres of the Tongass and Chugach National Forests to the Wilderness system. Currently, there are 33.2 million acres of National Forest System lands (17 percent of the total of 192 million acres) in the Wilderness system.

Recent Policy Developments

The Clinton Administration's Roadless Rule

In October 1999, the Forest Service initiated a Federal rulemaking process to put the remaining 58 million acres of unroaded National Forest lands more than 500 acres in size off limits to future roadbuilding except for specifically defined purposes. President Bill Clinton asked the Forest Service to begin an open public process to address how roadless areas within the National Forest system would be managed in the future. Roadless areas have typically remained roadless because of inaccessibility, rugged terrain, low timber values, or environmental concerns. The Forest Service Notice of Intent to the public cited budgetary concerns and public questioning of the logic of building new roads into roadless areas when the Forest Service receives insufficient funding to maintain its existing road system. The Forest Service has a growing $8.4 billion maintenance and reconstruction backlog and receives only 20 percent of the annual funding it needs to maintain its existing 380,000-mile road system to environmental and safety standards (USDA, U.S. Forest Service 1999).

After 23 public hearings held at locations around the country and a record 1,156,308 comments on the proposed rulemaking, a final Roadless Rule was issued in 2001 (USDA, U.S. Forest Service 2001). The rule had not yet been implemented by the succeeding George W. Bush Administration when it was set aside by a Federal district court in Idaho, which halted the rule's implementation on grounds that the process under which the rule was developed lacked adequate public participation. This decision was appealed to the Ninth Circuit Court of Appeals, where the case was decided in 2003 in favor of the Clinton Roadless Rule. At this writing, it is undecided if the state of Idaho will bring the matter before the U.S. Supreme Court.

Revision of National Forest Plans

Recommending undeveloped roadless areas on National Forests for permanent protection can also occur in the forest planning process under the National Forest Management Act of 1976. All 155 National Forests are in the process of or will embark on updating their forest plans, all of which were first developed in the 1980s. Although there is no guarantee that Congress will grant permanent protection to roadless areas on National Forests, the public involvement process in the forest plan revisions will likely keep the debate alive at local and regional levels. Forest plan revisions for all National Forests are scheduled to be completed by 2012.

Conclusion

This is quite likely the final chapter in the half-century struggle over wilderness protection on the National Forests. It is really part of a larger societal debate about the extent to which forests will be developed for commodity production and the extent to which forests will be protected to serve purposes such as biodiversity conservation that development would harm. There is broad support in American society for protection of remaining wild areas but a wide divergence of views over the questions of how much should be protected and where protection should take place.

Since the battle over wilderness was joined in the 1950s, some important changes have taken place that may have bearing on the outcome of this final chapter. Biodiversity conservation, and especially the protection of habitat for federally listed threatened and endangered (T&E) species, has become a leading issue in forest management in the United States on both public and private lands. Since 1989, sales of timber from the National Forests have declined from a high of 12 billion board feet annually to one-tenth of that. Small forest products companies—and the rural communities that depend on them as the linchpin of their local economies—have been hard hit. In many cases, communities have lost not only the forest products companies that have now gone out of business, but much of the economic infrastructure that had developed around these primary industries. For better or worse, it will be difficult to re-establish this infrastructure, even if National Forest timber sales once again begin to increase. This has now become a serious issue as the Forest Service attempts to thin forests and reduce hazardous fuels and can no longer find local firms capable of performing this kind of woods work.

Wilderness advocates point out that many of the remaining unroaded areas have been identified independently as areas of regionally or globally significant biodiversity values (WWF 2001) and that Wilderness protection would have little influence on most of the values for which these National Forest lands are currently being managed. Logging in designated Wilderness areas is

prohibited, but depending on the individual legislation that created them, some Wilderness areas still permit grazing and mineral development. Wilderness areas provide outdoor recreation
opportunities, preserve wildlife and fish habitat, and protect watersheds. From the standpoint of timber production, one of the
key reasons many of these areas are still unroaded is that it is un-
economic to develop the timber resource because of its remoteness or relative low productivity, or because of other factors like
steep terrain, unstable soils, or watersheds highly susceptible to
damage from soil erosion or landslides.

Whether or not most of this 58 million acres of remaining unroaded forest receives Wilderness designation may well be determined by the extent to which some level of forest management is
needed to prevent these areas from becoming sources of wildfires. Recognizing the essential role that fire plays in many forest
ecosystems, the Forest Service is more likely now than it was in
the past to allow lightning-caused fires in wild areas to burn. But
the Forest Service must take measures to minimize the likelihood
that fires will start in wild areas and then "blow up" to threaten
life and property outside the boundaries of Wilderness areas. This
is a tricky business that depends on many variables, such as future drought conditions and weather patterns, that are difficult to
predict with a high degree of confidence.

As in nearly every other aspect of natural resource management, decisions made on the basis of evaluations of site-specific
characteristics of individual roadless areas are likely to be superior than a single blanket decision made on the entire 58 million
acres. As a matter of politics, blanket land protections at such a
large scale are simply too contentious, and they invariably split
sides into polar-opposite positions. The legal challenge on which
the Federal district court in Idaho blocked implementation of the
roadless area conservation rule was not based on ecological or
even economic concerns, but the degree to which state and local
government and affected communities had been consulted in the
process. Wilderness designations following the Forest Service's
RARE II recommendations were handled by Congress on a state-
by-state basis, with a certain amount of deference given to the
consensus reached within each state by local government and the
state's Congressional delegation. It is likely that the future of the
58 million acres of unroaded areas on the National Forests will be
decided through a similar process.

Forest Health

In testimony to Congress in August 2002, Forest Service Chief Dale Bosworth reported that 73 million acres of the National Forests were at significant risk for outbreaks of catastrophic wildfires. This is an area more than 20 times the size of the state of Connecticut. Already that year, both Colorado and Arizona had experienced the largest wildfires in each state's history; these were followed by a wildfire in Oregon that now ranks as the largest in U.S. history. Federal fire suppression costs for the year exceeded $1.5 billion.

How does such a situation develop? In part, it can be attributed to the unusually hot, dry conditions throughout the western United States in that year, part of a drought that had affected the region for several previous years. But there is a larger issue—one that is a function of past and present forest management practices, the evolving scientific understanding of forest ecology, and incremental policymaking through the Federal appropriations process. In the nearly 100 years since the Forest Service began actively suppressing fires—effectively excluding fires from fire-adapted ecosystems—forests in many U.S. regions have transformed into dense stands with uncharacteristic mixes of tree species. Past logging and grazing practices have also contributed to what many regard as unhealthy forest conditions across large areas—conditions that are not resilient to disturbances like fire and are, in many cases, susceptible to uncharacteristically severe fires.

Background

To understand today's forest health situation, one must follow the clues back to 1910. It too was a year of unusually hot, dry conditions in forests in the West. In late summer, more than 1,700 wildfires exploded across Montana and Idaho, burning out of control and devouring everything in their path. The fires left more than 3 million acres of forest—about the size of the entire state of Connecticut—in charred ruin. Eighty-six lives were lost, and countless acres of private property and entire towns were destroyed. The 1910 fires also consumed an estimated 8 billion board feet of valuable timber, enough to build 450,000 single-family homes. Because wood was still a crucial raw material for virtually every aspect of American life, the fire was also economically devastating.

For years afterward, these fires had a profound effect on watersheds, wildlife habitat, and wood production in the region (see Pyne 2001 for a more detailed account of the 1910 fires).

The fledgling U.S. Forest Service, having been established and given charge of the National Forests only five years earlier, was itself devastated by this catastrophe. Subsequently, the agency undertook a dedicated effort to become an effective firefighting organization, developing many innovative tools and techniques based on extensive new research into forest fire behavior. The stated objective was to spot every fire and extinguish it within 24 hours; later the objective was changed to extinguishing every fire before 10 A.M. the next day—the so-called "Ten A.M. Rule." In effect, the Forest Service declared war on wildland forest fires, using heroic "smoke jumpers" and a fleet of bombers to dramatically attack fires from the air with chemical fire retardants. With its attention to organization and its responsive hierarchical command system, the U.S. Forest Service became one of the most admired and respected agencies in the Federal government. The agency mounted a public campaign to enlist every citizen in the fire-prevention effort; its mascot Smokey Bear, who admonished "Only You Can Prevent Forest Fires," remains today one of the most widely recognized symbols in advertising.

Ironically, the enormous success of this effort over many decades helped set up the dangerous situation we see in the nation's forests today. Only recently have scientists and forest managers come to understand the essential role that fire plays in many forest ecosystems. Some species, such as lodgepole pine, actually need fire to open their wax-sealed cones and release their seeds in order to regenerate. Western U.S. forests experience fires that periodically burn the underbrush, usually without killing the mature trees in the burned area. Many of these fires—far more than many realize—were ignited by Native Americans. Early European explorers and settlers encountered open, parklike forests with a relatively small number of large, mature trees and little woody vegetation in the understory. They most likely did not realize that Native Americans had maintained these areas using fire for thousands of years.

After the Forest Service had spent decades suppressing all fires, no matter the cause, understory vegetation in many of these forests became thick and dense, filled with tree species that historically had been absent. Forests in many parts of the western United States, instead of containing mostly pure ponderosa pine, now had mixed-conifer stands of trees, with Douglas fir, true firs, and juniper encroaching into these areas. This heavy vegetation fuels hotter and more persistent fires, often providing a lethal "ladder" to the towering crowns of the mature trees in the forest. The result is a forest that is killed entirely, often over a large area. With little to regenerate vegetation to provide ground cover, subsequent erosion by wind and water rob the soil of its nutrients. Unnaturally hot fires can cook forest soils until they become hardened and hydrophobic, meaning that precipitation tends to run off rather than being absorbed into the soil. The result is floods and negative effects on water quality and fish habitat.

Adding to the effects of overly successful fire suppression is the minimal level of reinvestment in many forest areas harvested in past decades. Following timber harvesting, reforestation occurs either through natural regeneration or through tree planting. If forest regeneration is to be successful—that is, if the tree seedlings are to effectively outcompete brush and other vegetation—there must be many more seedlings per acre than will later be found in a healthy mature forest. Some seedlings are eliminated by insects or disease, and others by competition with one another. But forests in which human actions such as large-scale timber harvesting have interfered with natural processes—particularly in fire-adapted ecosystems in which fire is artificially eliminated—some sort of management action is needed to take the place of these natural processes. Usually this involves thinning the stand, removing some trees and allowing others to continue growing with reduced competition for light, water, and nutrients. This action is often termed a *precommercial thinning,* because the trees removed are usually too small or poorly formed to be of any commercial value. Although this step is a key element in responsible forest management following timber harvesting, it is a significant cost that is borne years after the income from the prior timber harvest and years before the income from the next timber harvest.

In the case of the National Forests, the rate of timber harvesting was sharply increased during the postwar housing boom of the 1950s, but funding for reforestation and thinning did not keep pace with funding for timber sales. The cumulative result of a half-century of heavy cutting on the National Forests without the necessary reinvestment subsequent to timber harvesting contributed significantly to the current unhealthy conditions in many of these forests. Extensive areas that were clearcut in the 1940s and 1950s are now choked with densely packed trees. In the case of ponderosa pine, historically there were between 5 and 20 trees per acre because of frequent, low-intensity ground fires that killed young trees and shrubs. Today, these same areas carry up to 500 trees per acre or more. This has not only increased the risk of fire and pest outbreaks, but it has diminished the value of these forests for wildlife, watershed protection, recreation, and even wood production.

Recent Policy Developments

The increasing frequency and extent of destructive wildfires have made the forest health situation front-page news across the country. Wildfires have prompted a variety of responses from resource management agencies, legislators, the forest products industry, and environmentalists.

During the 1990s, resource management agencies stepped up their use of prescribed burning, or controlled fires, to reduce fuel loads and reestablish the natural role of fire in the ecosystem. Forest managers' use of controlled burning has been severely limited by concerns over the effect on air quality and by fears that managers would be unable to prevent the spread of fires into surrounding areas. Even with careful consideration of expected weather conditions at the time of a proposed controlled burn, forest managers often have not been able to obtain the necessary approval from the U.S. Environmental Protection Agency (EPA) or state environmental agencies concerned about particulate pollution and impacts on visibility. Public confidence in forest managers' ability to contain controlled burns has been damaged by several high-profile accidents, such as the National Park Service–prescribed fire in Bandelier National Monument in New Mexico, which got out of control and burned much of the town of Los Alamos in May 2000. Even "let-burn" policies (to allow lightning-caused fires in Wilderness areas to burn out naturally) have been undermined by cases such as the Canyon Creek fire that began in the Bob Marshall Wilderness but escaped to threaten the town of Lincoln, Montana.

The wildland forest fire problem has been exacerbated by an influx of people and homes into what is known as the wildland-urban interface. This is an area where human development of any kind has encroached into undeveloped lands. In the western United States, the fastest-growing areas in terms of population change are at the wildland-urban interface, specifically at the borders of Federal public lands. The 1985 Ojai fire in California was the first of many uncharacteristically severe wildfires occurring at the wildland-urban interface that destroyed property and threatened lives. The fact that the wildland-urban interface is quickly expanding further into wildlands and bordering on Federal public lands makes addressing wildfires and forest conditions even more challenging.

In September 2000, following a bad fire year in which a significant number of homes and other private property had fallen victim to forest fires originating on Federal public lands, land management agencies developed a strategy to systematically reduce fire hazards on both public lands and adjacent private lands. The result was the National Fire Plan, which gave priority to reducing hazardous fuel in forest areas within the wildland-urban interface and to making homes and other domestic structures less susceptible to fire damage through public education efforts conducted by the state forestry agencies. The National Fire Plan also emphasized community-based approaches through multiyear land stewardship contracts with local firms to do thinning on an ongoing basis.

Closely related to the emphasis on community-based approaches is the focus on developing and enhancing markets for so-called small-diameter wood that is usually removed from these thinnings. Traditionally, construction and similar sectors use larger timber, which retains its structural characteristics better than small-diameter timber. However, there are boundless opportunities—ranging from biofuels energy to particle board to window trim and sills—to transition from traditional wood products toward using materials made from small-diameter timber from thinning treatments. However, without adequate assurance that the timber supply will be available in the foreseeable future, few community entrepreneurs will be willing to risk investing in small-diameter wood. The National Fire Plan provided an Economic Action Program to jump-start these entrepreneurs.

A broad political consensus for the objectives of the National Fire Plan, including the forest products industry, the environmental community, and state governments, was developed by the Western Governors Association in 2002. Called the 10-Year Implementation Plan, the consensus signaled a commitment among Federal, state, local, and nongovernmental organizations to work together to address the twin issues of wildland forest fires and forest health.

However, before the National Fire Plan could have a major influence, forest fires in 2002 scorched more acres than in any year since the disastrous fires of 1910. President George W. Bush toured the burned-over forests and prompted the Healthy Forests Initiative to deal with what was regarded as a National Forest health crisis. Citing legal actions under national environmental laws like the National Environmental Policy Act (NEPA), the Endangered Species Act, and the National Forest Management Act as barriers to timely action to address forest health concerns, the White House initiative sought to streamline environmental review for projects to reduce hazardous fuels, to minimize opportunities for legal challenge, and to help finance the cost of forest health projects through the sale of commercial-quality timber in other areas of Federal forests.

One of the most controversial features of the Healthy Forests Initiative was to allow local Forest Service managers to designate virtually any fuels treatment project as a "Categorical Exclusion" (CE) under NEPA. A Categorical Exclusion exempts a Federal project from environmental impact analysis and documentation. Additionally, it precludes NEPA-based administrative appeals. Environmental groups fear that projects to remove large trees or even clearcut an area can be defined by the Forest Service as a fuels treatment and that such projects would automatically be excluded from any critical analysis and oversight. Environmental activists perceive this as a lack of accountability and are concerned that citizens will have no recourse concerning the fate of their public lands. From the perspective of others, the initiative will allow the Forest Service to take action on projects that have been held up by seemingly unending procedural delays, analysis, and appeals. At this writing several bills have been proposed in Congress to authorize some or all parts of the Healthy Forests Initiative.

Conclusion

From a forest policy perspective, the dangerous fire and forest health conditions prevailing in many areas of the nation's forests are in no one's best interest, and there is a broad consensus that something must be done to address these conditions. Despite our improved understanding of the natural role of fire in many forest ecosystems, restoring the natural fire regime is no longer an option because of the proximity of homes and other human development.

What is likely to emerge from all of this is a two-part strategy. The first part must address the near-term need to thin dense, overstocked forests, with priority given to inhabited areas in the wildland-urban interface. Dealing with this large backlog of deferred maintenance will be extremely costly and thus can be expected to take place over a period of several years. It has been suggested that this will require a systematic program of projects in rural areas throughout the nation's forests, not unlike the Civilian Conservation Corps that was successfully used to accomplish public works projects in the 1930s.

A second and longer-term strategy is needed for cost-effectively removing trees of various size and quality in areas where nearby communities make it impossible to let fire play its natural role. These communities have a disproportionate stake in maintaining the health of nearby forests for a variety of values and in preventing it from becoming a dangerous liability. Application of innovative, appropriately scaled technologies to improve the value that can be derived from processing small-diameter trees into manufactured products, biofuels, and cogenerated electricity can become the basis for stable, sustainable employment in rural communities and at the same time provide for continuous forest stewardship that will make today's dangerous forest health conditions a thing of the past.

Community Forestry

Forest policies affect communities in vastly different ways. Residents of a suburban neighborhood in Atlanta may support a policy that sets aside large amounts of National Forest lands in northwestern Montana for recreation and wildlife habitat; but for the residents of Libby in northwestern Montana, the same policy could eliminate jobs and tax revenues and, therefore, schools, local road maintenance and repair services, and other public services that are essential to the local community. The same forest that once sustained the community of Libby may now be the cause of its demise.

Community forestry focuses on the interdependence between healthy forest ecosystems and healthy social systems. Community forestry is a growing movement, both in the United States and abroad, yet at its roots it is a concept that is as old as the practice of forestry itself. Community forestry links efforts to develop new models of sustainable forestry with initiatives aimed at alleviating poverty in rural economies. It emphasizes decentralized natural resource management decision-making and promotes social and economic self-reliance on a relatively small scale. Community forestry in the United States today is largely a reaction to the political dysfunction of centralized decision-making in Federal forest management. But organizations favoring community-based approaches to forest stewardship also express growing concern over the consolidation of the global forest products industry into just a few national or transnational corporations controlling large expanses of forest land, the management of which is centrally determined by individuals who often have little or no knowledge of the local communities that are affected.

Background

Community forestry in the United States today is in many ways a rediscovery of the earliest concepts of forestry. The first glimmerings of what would later become known as sustained-yield forest management arose in communities throughout Europe in the late Middle Ages. Growing populations relied on forests for fuel, structural timber, wild game, forage, fodder, and bedding for livestock. Yet without a road system suitable for moving heavy goods such as wood for long distances, communities had to rely primarily on forests within easy reach for all of these goods and services—exporting and importing wood products was out of the question. To prevent these forests from being devastated through overuse and to avoid conflicts among users, these communities devised various approaches to governing the use of forests by individuals so as to ensure the well-being of the community as a whole and to keep forest utilization within the bounds of what could be sustained over the long term.

Early forest laws in colonial America reflected this same approach, but it was during the Conservation Movement of the late nineteenth century that community forestry found a place in national policy. Widespread "cut-and-run" forest exploitation had left many communities bereft of forests sufficient to meet basic needs for structural timber, wild game, or watershed cover to prevent catastrophic flooding. Forest Reserves were established beginning in 1891 to retain forests in common public ownership, provide an inexpensive local source of wood for developing communities, and prevent overuse by individuals for purposes such as timber harvesting and livestock grazing. At a time when the wealth of the United States was becoming increasingly concentrated in the hands of a few powerful individuals and corporations, the Forest Reserves were established to ensure that these essential resources remained available to the American people as a whole—and especially to the citizens of nearby communities.

Today there is a resurgence of interest in community forestry in the United States, particularly in connection with the Federal public lands. Following World War II, timber harvesting on the Federal forests more than quadrupled what it had been for much of the previous half-century as they became a primary source of timber supply for producers of commodity wood products. The increased timber supply provided a competitive advantage for larger corporate wood products companies. Increasingly, many rural communities found their economies turning on the fortunes of a handful of operations or even a single company. While the timber harvest remained high, everyone prospered. This prosperity largely ended in the early 1990s when concerns about declining environmental quality—and especially about habitat for threatened and endangered species—resulted in a sharp decline in timber sales from Federal lands. The dislocation of lumber and plywood manufacturing facilities in numerous rural communities resulted in economic hardship for many individuals who were unable or unwilling to relocate.

Since the early 1990s, a growing number of community-based forestry efforts have emerged to reconnect sustainable, healthy community development to sustainable, healthy forest ecosystems. Some early examples include the Applegate Partnership in southwest Oregon and the Quincy Library Group in northern California. Today, there are numerous community forestry efforts in every corner of the United States that include both Federal public land and private forest lands. National organizations such as American Forests, Forest Trust, the National Network of Forest Practitioners, and the Pinchot Institute for Conservation are good sources of more detailed information about the number, scope, and scale of community forestry efforts in the United States.

The sudden change in forest management practices also affected the forests themselves. Timber sales had long served as the mechanism by which the Federal agencies addressed other management needs, such as thinning dense forest stands, removing brush and other hazards, improving recreation areas, enhancing habitat for fish and wildlife, and maintaining many miles of un-surfaced roads. Before long, inattention to these basic land stewardship responsibilities was beginning to have negative effects, such as impacts on water quality and fish habitat caused by soil erosion from deteriorating roads perched on steep hillsides, often on unstable or highly erodible soils. Thickets of brush and unthinned stands of young trees created conditions ripe for insects and disease—and for wildfires—on many thousands of acres of Federal forest land. The declining condition of forests surrounding many communities, paired with the lack of Federal agency attention to these declines, have provided important niches that communities have identified and filled with varying success. For example, the National Fire Plan of 2000 and its subsequent 10-Year Implementation Plan emphasize community-based approaches to reducing wildfire risk and improving forest conditions.

Basic Elements of Community Forestry

Because of their dispersed, unique nature, community forestry efforts lack a unified framework. Indeed, community forestry efforts have emerged largely because unified policies and administrative frameworks have caused undesirable ecological, economic, and social situations locally. Nevertheless, four basic elements of community forestry have been synthesized by leaders in the field (Gray, Enzer, and Kusel 1998).

First, community forestry is about forest stewardship—maintaining and restoring the health of the land for current and future generations. Community forestry does not necessarily seek to maximize economic gains from timber only for the benefit of community residents. Maintaining and restoring forest ecosystems can provide a broad array of benefits and values to local communities above and beyond the financial returns from timber harvesting. These include but are not limited to watershed functioning, reduced fire risk, healthy populations of wildlife, unique biological diversity, and diverse recreation opportunities. In many ways, community forestry embodies multiple-use forestry at smaller geographic scales. The challenge to making community forest stewardship durable is the lack of financial investment opportunities for ecosystem maintenance and restoration. Aside from timber sales, hunting fees, recreation use fees, and water allocations, many goods, services, and values of forest ecosystems have no financial value, yet they contribute enormously to local communities. Linking ecosystem maintenance and restoration to the economic vitality of communities is often the biggest barrier to sustaining community forestry efforts.

Second, community forestry signals a shift toward more open, participatory decision-making processes. Currently, forest management decision-making, especially on Federal public lands, is dominated by organized interest groups that staked out certain policy and rhetorical positions regardless of the specific ecological and economic conditions affecting communities. The decision process is predictable: the Forest Service proposes an action; one set of organized groups supports the action; and another set of organized groups opposes the action. Communities are often left without a voice in the process. Community forestry efforts seek to overcome this positional approach to forest stewardship by developing processes "that recognize and respect the legitimacy of diverse perspectives, cultures, and knowledge systems" (Gray, Enzer, and Kusel 1998, 9).

Third, community forestry focuses on an indefinite time period—communities rarely have the choice of whether to stay or go. For a forest-dependent community to endure, it must ensure that the surrounding forest ecosystem is sustained into the future. This requires a firm commitment of monitoring and learning. Monitoring ecological, economic, and social conditions can provide information for ongoing planning and decision-making. Monitoring also ensures that local communities are accountable not only to themselves and their neighbors, but to nonlocal stakeholders, especially with respect to Federal public lands. From monitoring, communities learn about the interrelationships between their surrounding forests and their own communities. Learning is empowering, and communities can set off on the road to self-governance rather than waiting for someone else to make decisions that affect their fate and futures.

Lastly, community forestry in the United States is strengthening connections to a large global movement toward community forestry. With financial support from organizations such as the Ford Foundation, community forestry efforts in the United States have learned from and contributed to community forestry activities around the world. Community forestry was a central topic at the World Summit on Sustainable Development in Johannesburg, South Africa, in 2002, and U.S. community forestry practitioners were actively represented. A central theme in these discussions was the rapid globalization of forest products, fueled by large transnational corporations. Decisions about forest use, production, ownership, and conservation may not necessarily have forest-community interdependence in mind. The durability of community forestry in the United States and around the world depends on working through the tensions between global economic forces and local community priorities.

In sum, community forestry in the United States has filled a niche for forest ecosystem maintenance and restoration by emphasizing the link between sustainable, healthy forests and sustainable, healthy communities. Unique opportunities exist, primarily in the areas of wildland forest fire mitigation and watershed restoration. Each of these areas of ecosystem restoration falls outside the traditional timber focus that dominated forest-dependent community economic development. Instead of communities relying solely on traditional timber sales and revenues, they can generate economic benefits from a host of stewardship activities. The main challenge is how to make this link operational.

Recent Policy Developments

To address this situation, the Federal agencies have begun experimenting with new approaches to land management in closer partnership with local communities. The Forest Service, for example, has undertaken a nationwide pilot project in which the agency enters into contracts with local governments or private firms in local communities to provide a variety of land management services over a period of several years. These land stewardship contracts once again make local communities active partners in balancing the many uses of nearby forests and in ensuring the long-term sustainability of the wide variety of goods and services a well-managed forest is capable of providing. Through additional research and technical assistance, the Forest Service is helping communities to develop and market new products manufactured from the small-diameter trees being thinned from overstocked stands and to capitalize on high-value nontimber forest products (such as medicinal herbs, fir boughs, mushrooms, and floral ferns) that can be sustainably gathered from nearby forests. An example of this is the Economic Action Program under the National Fire Plan. Additional financial and technical resources are delivered through the Rural Community Assistance

Program of the Forest Service. The latter program is part of a broader effort under the U.S. Department of Agriculture. The ideal is local communities that contribute to the active stewardship of their surrounding forests and, in so doing, strengthen and diversify their economies on the basis of employment and income that can be sustained over the long term.

The need for community-based forest stewardship, particularly on Federal lands in the western United States, has gained added impetus because of several large wildfires in recent years—increasing the urgency to reduce buildup of fuels, salvage forest areas killed by insects or disease, and thin overstocked stands before they, too, become fire hazards. In the 106th session of Congress (1999–2000), a bill proposed by Senator Jeff Bingaman (D–New Mexico) entitled Community Forest Restoration Act was passed, granting funding and legislative authority to National Forests in New Mexico to work more closely with forest-dependent rural communities on forest restoration projects. In 2002, Senator Bingaman cosponsored a bill with Senator Larry Craig (R-Idaho) entitled Community-Based Forest and Public Lands Restoration Act, which sought to expand the New Mexico approach nationwide. Although the bills did not pass in the 107th Congress, no fewer than five bills relating to community forestry and forest restoration have been introduced in the 108th Congress (2003–2004), in addition to three appropriations bills granting funding to specific community-based forest restoration projects and programs. Given large wildfires' potential danger to life and property, many local communities adjacent to Federal forest lands have a direct stake in addressing forest conditions that increase the likelihood of a catastrophic wildfire. Mechanisms such as land stewardship contracts are likely to become essential tools to address the ongoing challenges of Federal forest land management—and to address them within the context of community forestry.

Because these are Federal lands managed for the benefit of the nation as a whole, there are national-level organizations that are wary of what they see as a shift to more local-level decisionmaking that might serve local interests in ways possibly detrimental to the larger national interest. There has always been some level of tension between national and local interests, particularly when the national interest was articulated primarily in a conservation context and local interests were seen as focused only on maintaining jobs in the near term. Decentralized decision-making has always been regarded as a hallmark (and one of the great strengths) of Federal agencies like the Forest Service. During the 1970s and 1980s when the high level of timber harvesting on Federal lands was being continuously challenged by conservation groups, decision-making in the agencies became increasingly centralized through legislation, through the annual appropriations process, and through planning decisions. But local communities have become so frustrated with being caught up in national issues and having only limited access to the Federal lands that surround them that they are actively exploring alternatives to federal management. The Forest Service is now striving to bring the pendulum back to the center and to restore effective decentralized decision-making within the bounds of national policy and law.

Multiparty monitoring could become an essential tool to provide independent assurance to the public that community-based stewardship and decentralized decision-making are indeed keeping within the spirit and letter of national-level law and policy. Multiparty monitoring, involving a diversity of interests, has been used successfully in many circumstances to evaluate agencies' performance and ensure accountability, especially in new stewardship contracting demonstration projects. Although it has been slow to be applied on Federal forest lands, independent third-party certification could be another useful tool for ensuring that local resource management decision-making is consistent with agreed-on standards and policies.

On the private-lands side of community forestry, a host of Federal and state policies and programs have been developed. Several programs under the 1996 and 2002 Farm Bills emphasize coordination among landowners before they receive financial and technical assistance to address issues like protecting wildlife habitat on private lands and improving water quality. The Environmental Quality Improvement Program is one such example. At the state level, numerous mechanisms facilitate coordination among private landowners, effectively creating a community-based approach to stewardship. In Colorado, for example, the Habitat Partnership Program administered by the Colorado Division of Wildlife provides financial and technical assistance to private ranchers to protect and enhance habitat on their lands for migrating wildlife. Although the focus is on rangelands, the connection to forest ecosystem stewardship is direct, because many wildlife species inhabit forested areas during certain parts of the season and migrate through or to private lands during other seasons. In many eastern states, the Coverts program explicitly focuses on cooperation and coordination among private forest landowners to protect and improve habitat for wildlife. The Coverts program is modeled after the Cooperative Extension Service's volunteer training program, in which active adopters of conservation techniques train their neighbors and peers.

Conclusion

Community-based forest management is an approach that involves local communities as an integral part of policy development and natural resource management decision-making. Participation by local communities in resource planning and management can both improve the effectiveness of forest management and help ensure that local communities benefit from the stewardship of these natural resources.

Public forestry agencies in countries worldwide are experimenting with various approaches to community-based forest management. In India, degraded areas of public forests are actually given to local communities, which provide the labor and protection to allow the forests to regenerate in return for access for fuel, food gathering, and eventually a share of the returns from timber production. In parts of Africa, local communities are given a custodial interest in forest land to ensure that it does not become degraded. Local communities become stakeholders of the forest, not just users of its products.

In the United States, forestry agencies and conservation interests are recognizing that local communities have a strong interest in forest stewardship that maintains environmental quality, aesthetic values, and recreational resources—as well as sustainable levels of wood production—as the basis for a resilient, diversified local economy and a desirable quality of life. As Federal and state forestry agencies continue to restructure and downsize, they will increasingly turn to local communities to assist them in protecting and managing public forests. It is likely that national-level organizations representing both conservation and resource development interests will continue to seek refinements in the overall policy framework guiding the management of Federal lands to ensure that management decisions based on community-level consensus continue to reflect the broader public interests in these lands.

Forestry Regulation on Private Lands

Few topics are as controversial in U.S. forest policy as government regulation of private forest land use and management practices. The controversy dates back to the early 1900s, when Gifford Pinchot, the first Chief of the Forest Service, emerged as a strong advocate of federal regulation of private forestry. Although Pinchot's vision of federal regulation had little political support, regulatory policies and programs at the state level grew in number throughout the twentieth century. Today, 38 states have a law or regulatory program concerning private forestry practices. Some state laws concerning forest practices are comprehensive and rival Federal policies in their scope and breadth. Other states have restricted their role in regulating private forestry practices to the protection of water quality. The diverse regulatory landscape in U.S. forest policy is a product of evolution and change over the past 60 years. During this time, there have been three "generations" of forest-practices regulation.

Background

State regulation of forest practices on private lands has developed by stages. Starting in the early part of the twentieth century, the first generation of state laws governing forestry focused largely on the reforestation of cut-over lands, halting the cut-and-run approach that had previously characterized so much of timber harvesting in the United States. The second generation of state laws, starting a half-century later, focused on minimizing the effects of forest management practices on environmental quality, with a particular emphasis on protecting water quality. The third generation of such laws, starting in the last decade of the twentieth century, focused on protecting a broader set of public values, such as habitat for endangered species and public watersheds.

First Generation: 1920–1950

The primary concern in Pinchot's day was an impending "timber famine"—a shortage of timber supplies caused by cut-and-run logging practices. More often than not, lumber companies would not replant trees on lands they had cut over. Instead, these lands would become thick with brush or "trash trees"—tree species that had no merchantable value. Meanwhile, wood was a primary material for industrial, manufacturing, and residential development for a growing nation. The increasing demand for wood paired with what was perceived as a dwindling supply of available domestic timber spurred the call for greater government control over private forestry.

By 1920, it was clear that the Federal government would not have a role in regulating private forestry practices. Instead, the burden of ensuring socially beneficial forestry practices fell on the states. A handful of states with significant pine forests enacted what were called *seed tree laws*—statutes that mandated the retention of mature pine trees after harvest to ensure a seed source for the next generation of forest. An example is Louisiana's Turpentine Seed Tree Law of 1922. The fear of a timber famine reached its height during World War II, when timber was used in all aspects of the war effort. During the 1940s, many states passed laws that required private forest landowners to immediately replant their cut-over lands and use logging methods that would not adversely impact soil productivity (see Table 2.1).

Central to the controversy over government regulation of private forest land use and management practices is the "takings" doctrine—when government regulation limits uses or actions on private lands for public benefit without compensating the landowner for the lost opportunities. Takings is a clause in the Fifth Amendment of the U.S. Constitution that states: "No person

Table 2.1   States with Early Private Forestry Regulations

State	Year
California	1945
Florida	1943
Idaho	1937
Maryland	1943
Massachusetts	1943
Mississippi	1944
Missouri	1945
Nevada	1955
New Hampshire	1949
New Mexico	1939
New York	1946
Oregon	1941
Vermont	1945
Virginia	1950
Washington	1945

Source: P .V. Ellefson and F. W. Cubbage. "State Forest Practices Laws and Regulations: A Review and Case Study for Minnesota." *Station Bulletin 536-1980.* St. Paul: Minnesota Agricultural Experiment Station, University of Minnesota, pp. 9–10.

shall . . . be deprived of life, liberty, and property without due process of law; nor shall private property be taken for public use without just compensation." In 1887, the U.S. Supreme Court upheld government restrictions on private land uses in *Mugler v. State of Kansas* (123 U.S. 623, 1887), setting precedence in favor of government's police power to regulate if it is done to protect public health, safety, and morals (or, in the language of the Court, to avoid a "nuisance"). The landmark ruling in *Village of Euclid v. Ambler Realty Company* (272 U.S. 365, 1926) solidified government's police power over private land use and management practices that are deemed contrary to the public's welfare.

The takings issue on private forest lands reached a head in the late 1940s. The spate of state reforestation laws and regulations in the 1940s imposed increased costs to private forest landowners. Many owners were of the opinion that these state laws amounted to a taking of private forest land property without just compensation. The first major forestry-related takings court case was filed in Washington state by a landowner named Avery Dexter who, between 1945 and 1947, harvested 150,000 board feet of timber from his 320-acre parcel in northeastern Washington. In 1947, the Washington state forester directed Dexter to cease operations until the necessary permits were secured in compliance with the state's Forest Conservation Act. When Dexter failed to apply, the state halted further operations. In the lawsuit, Dexter made four claims:

- Washington's Forest Conservation Act permits what amounts to a taking of private property without just compensation.
- The act establishes an unreasonable exercise of police power.
- The act violates private property rights and impairs the obligation of contracts, which are guaranteed by the U.S. Constitution.
- The requirements stipulated in the act are unreasonable, unnecessary, arbitrary, and oppressive.

The case was first handled by a state district court, which ruled in favor of Dexter and found the Forest Conservation Act unconstitutional. The ruling was held as a victory for private forest landowners around the United States. But the state appeals court overturned the district court's ruling, leading to a showdown in the state supreme court, which ruled in favor of the state, asserting that "the state is not required by the United States Constitution to stand idly by while its natural resources are depleted, as affecting validity of conservation measures . . . [and that the] protection and conservation of natural resources constitutes a reasonable exercise of police power" (*State of Washington v. Avery Dexter* 1949). Perhaps the most compelling point in the Dexter case was made by Justice Matthew W. Hill of the state supreme court:

> Rights of property, like all other social and conventional rights, are subject to such reasonable limitations in their enjoyment, as shall prevent them from being injurious, and to such reasonable restraints and regulations established by law, as the legislature, under the governing and controlling power vested in them by the constitution, may think necessary and expedient.

The Washington supreme court's ruling was upheld on appeal by the U.S. Supreme Court without comment (338 U.S. 863 1949). As such, the Dexter case set judicial precedent regarding the constitutionality of a state's police power to protect the state's forest resources via regulation.

In general, strict reforestation regulations were broadly supported, especially in states like Washington and Oregon that are economically dependent on forest products. Many in the forest products industry were already active in reforestation, with the understanding that the future of the industry was directly tied to successful replanting of cut-over areas. The laws were also popular with the public, which sought to put an end to cut-and-run logging practices. However, this first generation of state-level forest practices laws opened the door for continued expansion of state police power over private forest land use and management.

Second Generation: 1970–1983

The 1970s brought significant changes in environmental policy at the national level. The National Environmental Policy Act of 1970 made the Federal government directly accountable for its effect on the environment. The Endangered Species Act of 1973 signaled growing public value of all living things, not just those that provide economic benefit to humans. The clearcutting controversy on National Forests was in full swing, capped by the seminal court case *Izaak Walton League of West Virginia v. Butz*, in which the Federal appeals court ruled against the Forest Service for violating the Organic Administration Act of 1897. Congress enacted the

National Forest Management Act of 1976 in response to the clearcutting controversy.

Public concern over the environmental impact of logging spilled over into private forestry. Starting with the Oregon Forest Practices Act of 1971 and ending with the revised Massachusetts Forest Cutting Practices Act of 1983, eight states enacted comprehensive regulatory laws and programs (see Table 2.2). Without a doubt, the major force behind the development of comprehensive state-level forest practice acts was the Federal Water Pollution Control Act of 1972. Public pressure and political action led to the passage of the act, which targeted both point and nonpoint sources of water pollution. Water pollution was a high-profile public concern at this time, especially after the Cuyahoga River caught fire in June 1969 because of high concentrations of pollutants.

Point sources are identifiable discharges of pollution, like outflow pipes from factories. The Federal Water Pollution Control Act of 1972 achieved a high degree of success in bringing point pollution sources under regulatory compliance. Nonpoint sources of water pollution are dispersed and harder to identify accurately in terms of their contribution to water pollution. Examples of nonpoint sources are agricultural operations, parking lots, construction sites, residential lawns, roadways, and golf courses. Forest practices such as road building, logging, skidding, and site preparation (for example, burning and fertilization) are also considered nonpoint sources of water pollution.

Nonpoint sources are more difficult for a centralized authority like the U.S. Environmental Protection Agency to regulate. Instead of letting the Federal government take the lead, the Federal Water Pollution Control Act of 1972 delegated responsibility to

Table 2.2   States with Comprehensive Private Forestry Regulations, 1971–1983

State	Year
Alaska	1978
California	1973
Idaho	1974
Massachusetts	1983
Nevada	1971
New Mexico	1978
Oregon	1971
Washington	1974

Source: P. V. Ellefson, A. S. Cheng, and R. J. Moulton. "Regulation of Private Forestry Practices by State Governments." *Station Bulletin 605-1995.* St. Paul: Minnesota Agricultural Experiment Station, University of Minnesota, p. 25.

the states. Specifically, section 208 of the act directed states to assess nonpoint sources of pollution and develop best management practices for major categories of land use to manage those nonpoint sources. States with active private forestry operations were increasingly pressured by the EPA and state-level environmental organizations to develop mandatory forest practices standards. Fearing federal regulation of private forest practices, timber industry trade associations in these states took the initiative and secured passage of forest practice acts. The timber trade associations in Oregon were the most proactive, working with that state's legislature even before the Federal Water Pollution Control Act was enacted to win passage of the Oregon Forest Practices Act of 1971. The Oregon FPA strengthened the reforestation standards of the Oregon Forest Conservation Act of 1945 and went further, imposing a more comprehensive regulatory process.

In general, state-level forest practices acts (FPAs) follow a common model, although details differ from state to state. The FPA model contains the following basic components:

- Rulemaking authority: delegation of authority to a board or the state forestry or natural resource agency to develop and implement administrative rules to carry out the FPA.
- Administrative procedures: notification and application process, requirements for written plans, administrative review protocols, provisions for variances or exemptions, and on-site inspections.
- Timber harvesting standards: limits on clearcut size and timing; recommended harvesting methods for steep slopes; guidelines for slash and debris disposal; and minimum standards for felling trees, cutting timber to more manageable lengths, and transporting logs from the logging site.
- Streamside restrictions: buffer areas or setbacks within which harvest activities are limited, restrictions on timber harvesting methods, and guidelines for machine operations in streamside buffer areas.
- Reforestation standards: minimum stocking standards (number of trees per acre planted) according to site quality and regional conditions, recommended regeneration methods by commercial tree species, site preparation standards, restrictions on chemical use, inspections, and minimum free-to-grow standards

(minimum height and vigor of trees five years after harvesting).

- Sensitive resource standards: special rules pertaining to protection of streamside areas, endangered fish and wildlife habitats, areas with steep slopes or unstable soils, and wetlands.
- Enforcement and penalties: enforcement procedures, notification of violation, stop-work orders, injunctions, damage repair orders, liens on property, fines, penalties, and appeals.

Because Oregon's FPA is regarded as a prototype law and program, it is worth exploring its features in greater detail. The Oregon FPA established the Oregon Board of Forestry, a quasi-legislative body composed of citizens nominated by the governor and approved by the state senate. The Board of Forestry had the power to draft and amend the administrative regulations that implemented the Oregon FPA. The Board of Forestry also became a focal point for public involvement. Board meetings were open to the public, something that was previously unheard of. That the public had a voice in determining private land use and management practices was more than unsettling to many, especially to professional foresters. However, the Oregon FPA reflected growing environmental concern among the public and reflected national trends.

Oregon's FPA also delegated authority for monitoring and enforcement to the Oregon Department of Forestry. To implement and enforce the Oregon FPA standards and associated regulations, the department created a force of forest practices officers whose sole duties were to review proposals of private forest landowners, monitor compliance, and punish violations. The regulatory responsibilities of Oregon Department of Forestry staff were a marked change from what they had done previously, which was to work with private forest landowners and timber operators in an educational or technical-assistance capacity. The Oregon FPA became—and remains—the model state-level forest practices act for states to follow.

For the private forest landowner, the Oregon FPA lays out specific mandatory guidelines. The first is a notification and approval process that varies according to potential environmental impact. A landowner or operator who contemplates any forest practice on private land is required to submit a notice of intent to the Oregon Department of Forestry. Depending on the proposed forest practices, operators are required to submit a written plan that identifies environmental impacts and describes how those impacts will be minimized. Forest practice operations are classified by the following criteria: proximity to sensitive watercourses and water bodies, size and location of clearcuts, proximity to visually sensitive corridors, location of sensitive wildlife habitat, and risk potential related to steep slopes and fragile soils.

After receiving approval from the Oregon Department of Forestry, the private forest landowner or operator must carry out the operations in compliance with the following regulatory standards:

- Clearcuts must not exceed 120 acres within a single ownership and must be at least 300 feet away from other clearcuts unless prior approval is given.
- Harvested areas are considered under compliance if reforestation meets or exceeds 200 seedlings per acre that have survived beyond 48 months.
- Forest practice operations are prohibited or restricted in streamside buffer zones and vary in specific standards according to whether or not the stream is fish-bearing.
- In clearcuts exceeding 10 acres in size, operators must leave at least two standing dead trees ("snags") and two standing live trees per acre that exceed 30 feet in height and are 11 inches in diameter. At least 50 percent of the standing live trees left in the harvested area must be conifers. In addition, two downed logs must be left per acre that must be at least 16 feet long and 12 inches in diameter. At least 50 percent of the downed logs left in the harvested area must be conifers.
- For specified scenic state and interstate highways, visual buffers must be retained 150 feet from the outermost boundary of the right-of-way. Within each visual buffer area, at least 50 healthy trees of at least 11 inches in diameter must be temporarily left per acre. Harvest debris must be removed within 30 days of completing the operation. Reforestation of at least 400 trees per acre must be done.

The notification and approval process and specific forest practice standards are not limited to the Oregon FPA. Other state FPAs have similar guidelines; California has the most comprehensive set of rules among all the states. In California, private forest landowners are required to develop a timber harvest plan approved by a state-licensed forester. The written plan must also undergo review and approval by the California Department of Forestry officials. In Washington, large private forest landowners are required to complete a cumulative effects analysis—a comprehensive study of the effects of forest practices on a range of ecological values over time. This is similar to the cumulative effects analysis required under the National Environmental Policy Act.

The administrative structure of forest practices programs also varies among states. Alaska, California, and Washington, like Oregon, have quasi-legislative boards that develop and modify forest practices standards and rules. In Idaho, Massachusetts, Nevada, and New Mexico, the state forestry agency has sole authority to develop, implement, and enforce standards and rules. Active enforcement and monitoring of private forest practices varies widely among states, depending on available staffing and budgets as well as the attitudes of agency directors and staff. Since initial passage, the state-level FPAs and administrative programs found in the aforementioned seven states have undergone varying degrees of revision and amendment. The most active states are California, Oregon, and Washington.

Third Generation: 1984–1995

State-level regulatory laws and programs expanded significantly in the 1980s due to three major factors: persistent water quality concerns and passage of the Clean Water Act Amendments of 1987, the proliferation of local ordinances, and persistent concerns over the environmental impacts of timber harvesting.

Concerns over Water Quality

Nonpoint-source water pollution remained a high-profile environmental policy issue at both the federal and state levels throughout the 1980s. States varied widely in their compliance with section 208 of the Federal Water Pollution Control Act of 1972. A national assessment of CWA implementation showed a lack of significant improvement nationwide. In 1987, Congress amended the Federal Water Pollution Control Act through the Water Quality Act, which required states to perform comprehensive assessments of waters that were significantly degraded from nonpoint-source pollution. Section 319 of the 1987 act required states to specify both voluntary and regulatory programs to deal with nonpoint sources.

The persistent concern over water quality at the federal level prompted action at the state level. Several states, including Florida, Maryland, Montana, New Hampshire, North Carolina, Vermont, and Virginia, included forestry under the umbrella of nonpoint sources to be regulated. Although the details of each state's regulatory program vary, in general all programs use a graduated sanctioning system. If forest practices on private lands conform to best management practices (BMPs) as defined by the state forestry or natural resource agency, then private forest landowners are not subject to any further requirements. However, if, on inspection, forest practices violate BMPs, the state forestry agency is empowered to take a series of steps to ensure compliance, starting with a notice of violation and ending with stop-work order, fines, and damage repair orders to be paid in full by the private forest landowner or operator. In Florida and Maryland, forest practices that affect wetlands are required to file for permits and may even require a written management plan.

The graduated sanctioning system is referred to as "contingent regulation"; regulatory enforcement comes into play only when voluntary BMPs are not followed and notices of violation are ignored. In these states, it is more politically palatable to leave regulation as the last resort. Regulation is invoked only to punish the "bad apples" in an otherwise professional forestry community. Contingent regulation programs are also favored by state forestry and natural resource agencies, because they retain their role as service foresters rather than police officers. Besides the aforementioned states, many others have developed the statutory authority to go after landowners or timber operators who egregiously impact water quality as a result of forestry practices.

Proliferation of Local Ordinances

State-level forest practices laws and programs also expanded in the 1980s and early 1990s in response to an increasing number of local ordinances. Such ordinance would change from municipality to municipality and from county to county. For example, 11 counties in Georgia have logging ordinances; 100 municipalities in Illinois have forestry-related ordinances; 20 of the 23 counties in Maryland limit forestry practices; 300 municipalities in New Jersey regulate forest practices; and municipalities in New York, Pennsylvania, Vermont, and Wisconsin have ordinances pertaining to logging or other forest management activities. State-level regulations offer a degree of consistency, especially for forestry operators.

A prime example of a state that expanded its private forestry practices programs is Maryland. The Chesapeake Bay Critical Areas Protection Act of 1984 provided broad authority to regulate land use activities to minimize nonpoint-source pollution to Chesapeake Bay. The state developed a regulatory program that requires private forest landowners to prepare timber harvest plans and sediment control plans and to maintain buffer areas for water quality and wildlife habitat protection. The regulatory program also establishes minimum standards for maintaining and enhancing forested vegetation in critical areas. The Chesapeake Bay protection act supplements a suite of state-level laws that regulate timber harvesting on private forest lands.

Public Concerns over Environmental Impacts

The association between environmentally harmful forest practices and the forestry profession has made it necessary for the forestry profession to become more proactive. Throughout the mid-1980s and early 1990s, state legislatures—prompted by environmental organizations, state forestry and natural resource agencies, forest products companies, and private forest landowners—addressed public concerns about the environmental impacts of forestry practices through a variety of mechanisms.

One approach is the forest practices act, which is similar to those found in the western states but not as comprehensive. Maine enacted the Maine Forest Practices Act in 1989 in response to public concerns over clearcutting and rapid development of mill facilities. The Maine FPA requires private forest landowners and operators to submit a notification of intent to harvest to the Maine State Forest Service for review and approval. The Commissioner of Conservation develops rules and standards, with primary focus on clearcutting restrictions and reforestation requirements. Clearcuts over 50 acres must have an approved management plan and are subject to inspection. Forestry practices on private lands in Maine are also subject to guidelines to protect water quality, soil, and wildlife habitat.

A second approach is to require registration and licensing of forestry professionals, operators, and forest product companies. The Connecticut Forest Practices Act of 1991, in addition to setting minimum standards for timber harvesting, establishes an examination and certification process for forestry professionals and operators. The underlying premise of the Connecticut FPA is that accountability for private forestry practices starts and ends with professional field foresters and operators. Through a certification process, the state and the public are able to track the performance of forestry professionals and operators. Violation of certification requirements or minimum forest practices standards could result in the revocation of one's license by the state's Commissioner of Environmental Protection. The state of West Virginia has similar requirements under its program titled Sediment Control during Commercial Timber Harvesting Operations. Persons engaged in timber operations, buying, and selling must be licensed by the West Virginia Division of Forestry. The division must also be notified in advance of intent to harvest timber, which includes how harvest operations will comply with best management practices.

Conclusion

Regulation is a useful policy tool to enforce minimum standards for private forestry practices. However, it has limited value as an incentive for private landowners and timber operators to go above and beyond those minimum standards. State forestry agencies generally regard regulation as a necessary backstop, one that serves as one tool in the toolbox. States with comprehensive FPAs continue expanding the scope of the laws and programs. One trend is to convert forest practices rules and standards into statutory provisions. What this does is to remove the flexibility with which rules and standards are applied in the field. The landowner or operator must comply no matter the uniqueness of the circumstance. This trend toward more "prescriptive" standards opposed to "discretionary" standards ensures consistency across all forest practices.

States without comprehensive FPAs but with some form of regulation of private forestry practices continue to experiment with the appropriate mix of nonregulatory and regulatory approaches. The development of contingent regulations seems to be a growth area. In general, state forestry agency staff would rather work with private forest landowners and operators through education and technical assistance programs than as enforcers of regulatory standards. Senior administrators in state forestry agencies contend that nonregulatory approaches preserve working relationships with landowners, an important consideration in managing a long-term resource like forests.

Minnesota provides a unique model for other states to explore in the future. In 1990, under pressure from public interest groups concerned with the proposed expansion of chip mills and, thus, increased harvesting of aspen-dominated forests, Minnesota's state legislature authorized a statewide environmental impact statement (EIS) to explore potential affects of alternative harvesting scenarios. It was the first statewide forest EIS of its kind in the United States. The EIS examined the effects of different harvesting levels on a broad range of forest-related resources, from wildlife and water quality to recreation and tourism. The policy outcome of the EIS was the creation of the Minnesota Forest Resource Council (FRC), a diverse group of citizens and agency leaders appointed by the governor to set broad statewide forest policy priorities. Site-specific best management practices and monitoring programs were developed at a regional level but were consistent with broad statewide priorities. The FRC convenes on a regular basis to continually review and revise statewide priorities and site-specific programs. This ongoing adaptive management approach stops short of outright regulation but is far more proactive than traditional education and technical assistance programs.

Finally, many of the changes in state-level regulation of private forestry practices are spurred by forces beyond government. Many states including California, Maine, and Oregon have relatively easy requirements for creating ballot initiatives, which are proposals put forth by nongovernmental actors (individual citizens or interest groups) that are voted on in general elections. All it takes to get an initiative on the electoral ballot is a certain number of petition signatures—88,000 signatures in Oregon. Since 1980, California, Maine, and Oregon voters have considered ballot initiatives to halt the use of clearcutting as a timber harvesting method. The campaigns to support or oppose ballot initiatives are fierce and highly charged. Thus far, voters in all three states have voted down these initiatives. However, citizens and groups strongly opposed to clearcutting continue their efforts, and it is a sure bet that initiatives to ban clearcutting will continue to appear in the future.

Forest Land Conservation

The single greatest threat to sustainable forest management in the United States is the loss of the forests themselves to development or conversion to other nonforest land uses. Most of the forest policy issues addressed in this book deal with the debates over how forests should be managed and under what standards or practices. While these debates are taking place, however, millions of acres of U.S. forest land are lost each year to other uses through development that—at least in terms of human time scales—is essentially permanent. Although some of this forest land development is appropriate and can reasonably be expected with increases in population, in many other instances forest land conversion is an unintended consequence of policies established for other unrelated purposes, such as taxation of property or inheritance. Unless these unintended consequences are recognized and addressed, and policies developed that support rather than discourage forest ownership and responsible stewardship, we can expect to see a continued loss of forest land, along with the important public values and ecological services these lands provide.

Conservation of Public Forest Lands

As noted previously, the idea of holding forest land as such and investing in its long-term productivity and sustainable management is a relatively new idea in the United States. Hardly more than a century ago, it was a widely accepted practice to acquire a tract of forest land, strip it of all valuable timber, and then sell it for agriculture or simply abandon it and move on to the next tract. The consequences of this practice, in terms of future timber supply—but also disastrous floods and catastrophic forest fires that caused widespread destruction of lives and property—first inspired the idea of reserving large areas of forest land in public ownership for public purposes. One of the first such reservations was the Adirondack Park, a 4-million-acre area in upstate New York that was protected by the state legislature in 1892.

The Forest Reserve Act of 1891 authorized the President to reserve Federal forest lands from the public domain by executive order, but the act said nothing about how these lands would be managed or by whom. The Transfer Act of 1905 placed these Forest Reserves, which were subsequently renamed National Forests, under the stewardship of the U.S. Forest Service in the Department of Agriculture. The agency's first Chief Forester, Gifford Pinchot, and President Theodore Roosevelt greatly expanded the National Forest System, which now encompasses 192 million acres in the continental United States, Alaska, and Puerto Rico. History may show that Gifford Pinchot's greatest contribution as the nation's first native-born forester was not his introduction of scientific forestry in the United States. Rather, it was his role in establishing the National Forests and conserving them in public ownership to serve a variety of evolving public needs that could not even have been guessed at in Pinchot's time and that we cannot predict for the future.

Additional reservations of Federal public lands continue to be made, primarily through the Land and Water Conservation Fund (LWCF). A portion of the Federal revenues from oil and gas production on the Outer Continental Shelf are designated each year for acquiring lands of significant conservation value. Of these designated revenues, however, only those specifically authorized by Congress during the annual appropriations process can actually be used for land acquisition (the remainder accumulates as a "balance" that increases over time but, like Social Security trust funds, this balance is only on paper and does not represent a large cash reserve that can be tapped at any time). In recent years, the authorization of LWCF expenditures has become an ideological battleground in Congress over the wisdom of adding to the portion of land under federal ownership.

This ideological battle has extended to include various proposals to divest large areas of Federal lands through privatization. Especially in states west of the Mississippi River, large areas of land were reserved to the public domain as a condition for statehood, and many of these lands remain under federal ownership and management. Although most of these lands remain open to grazing, mining, timber harvesting, and other uses, these uses are regulated by one or more Federal agencies, which has created resentment among individuals who feel these lands could be better managed under private ownership. The "Sagebrush Rebellion" of the 1980s and the States' Rights, County Supremacy, and Wise Use movements in the 1990s were examples of efforts over the years to privatize Federal lands, but they failed to gain the necessary support at the national level. The political resistance to efforts to transfer authority over Federal lands to individual states is often attributed to the disproportionate political power in urban areas, especially in the East, where there is a strong interest in continued federal management of these lands to protect water supply, prevent overuse of resources, and conserve environmental values.

However, careful economic analysis often shows that federal transfer payments to states with a high percentage of Federal land significantly exceed the income the state could expect if the lands were in private ownership. Moreover, if the Federal lands were offered for sale, the highest bidder would in many cases not be local ranchers and other local interests who currently are the major users of these lands—and local users could find themselves either shut out or paying far more to the new private landlord than they have become accustomed to paying for Federal use permits. To address these concerns, there have been related proposals to divest Federal lands into ownership and management by state governments. Although this option might offer a greater sense of local control, the economics are even worse from a state government perspective: not only would the state lose the value of the federal transfer payments, but it would also have to assume the responsibilities and costs of management. Nevertheless, the notion of increased local control over lands whose management can significantly influence the course of local economic development remains powerful at the visceral level, and successfully addressing both the local and national interests will continue to be one of the most important challenges facing Federal land management agencies—and thus the continued conservation and stewardship of these forests as Federal public lands.

Conservation of Corporate Forest Lands

In spite of the attention focused on forest management issues on public lands, it is important to remember that three-quarters of U.S. forest land remains in private ownership. Furthermore, the ownership and management of private forest lands in the United States is currently undergoing greater change than at any time in recent history. This will have important implications for the continued conservation of these lands as forests and for the important public values these lands provide simply stemming from the fact that they are forests rather than suburban subdivisions or strip malls.

About one-fifth of private forest land in the United States is owned by integrated forest products companies, which produce lumber, panels, paper, and paperboard. Historically, the other four-fifths of private forest land has been owned by "nonindustrial private forest" owners, a wide variety that includes everything from farmers to Native American tribes. However, powerful economic and political forces are substantially redrawing the picture of who owns America's private forest lands and how they are managing them.

The forest products industry is one of many industries that have been significantly reconfigured in recent years, first by the trend toward consolidation at the national level and then by economic globalization. The spasm of mergers and acquisitions beginning in the 1980s and continuing early in the twenty-first century has resulted in an unprecedented consolidation of hundreds of small and medium-sized forest products companies into an industry now dominated by fewer than a dozen large companies worldwide. The top five—International Paper, Georgia Pacific, Weyerhaeuser, Stora Enso, and Smurfit Stone Container—produce 20 percent of the world's industrial wood. In some cases, companies were acquired by speculative investors who determined that the market value of a company's assets was greater than the combined value of its outstanding stock. The investors made "tender offers" to stockholders to purchase all their shares at a price higher than the value at which the shares were currently being traded. Having acquired a controlling interest, the investors then proceeded to break the company up into its component parts, which were then sold off separately for a tidy profit. Because a company's forest lands are shown on its books at their purchase price rather than at a current appraisal of the value of the land and timber, these forest lands represented "undervalued assets," and those companies with large timberland ownerships were especially vulnerable to such hostile takeovers.

A few high-profile instances of large tracts of forest industry land being sold to real estate developers has prompted some interesting responses. Concern over several such sales in northern New England and New York resulted in Congress creating the Forest Legacy program, which provides funds to state agencies to acquire ownership or conservation easements on forest lands that are under imminent threat of development. Such concerns also prompted a significant increase in the activity of private conservation organizations such as the Nature Conservancy, the Conservation Fund, and the Trust for Public Lands, which could act quickly to acquire forest lands as they came on the market and then transfer the lands into public ownership as soon as public funds became available.

But the continuing consolidation of the forest products industry and a series of divestitures of hundreds of thousands of acres of forest land that are no longer strategically valuable to the new combined companies has overwhelmed land conservation approaches that have worked well in the past and stimulated the development of new approaches and institutions. Land trusts and nongovernmental organizations (NGOs) that previously functioned primarily to purchase and resell forest land into public ownership have now become major owners and managers of forest land in their own right. The Nature Conservancy, for example, now owns 15 million acres in the United States that it manages as a system of private preserves.

Timber investment management organizations (TIMOs) have emerged as a new category of private forest landowner, raising capital primarily from pension funds and other institutional investors to respond to major acquisition opportunities as they arise. TIMOs are able to offer substantial return on investment through both current income from forest management and appreciation of land and timber values over the period of the investment. TIMOs are not land conservation organizations, and at the end of the investment period they usually sell to the highest bidder to maximize return to investors. However, with the access to capital that enables them to act quickly to acquire large tracts of forest land being divested by forest products companies, TIMOs can buy time for conservation organizations and public agencies to carefully assess and set priorities concerning the lands they will eventually endeavor to acquire from the TIMOs.

The large-scale forest land divestitures have also forced innovation in the traditional use of conservation easements to conserve forest land and stave off development. The opportunities and needs for conservation easements have almost overwhelmed all possible sources of funding for acquiring them. "Working forest" conservation easements have emerged as means of explicitly encouraging active forest management, in accordance with specified standards of responsible forestry, to help offset the cost of acquiring an easement aimed at preventing land conversion and development. Congress is currently considering proposals to make it easier for NGOs to raise capital for private land conservation and working forest easements by altering Federal income tax policy to authorize "conservation bonds," which would be exempt from Federal income tax, much like tax-exempt municipal bonds.

Conservation of Family Forest Lands

Few issues in U.S. forest policy have been debated longer—and less productively—than how to best ensure conservation and responsible forest management on the nonindustrial private forests that collectively constitute three-fifths of the nation's forest land.

The right of private individuals to own land with secure tenure and a high degree of self-determination in the use of that land is one of the most basic tenets of the U.S. Constitution. Certainty that these rights will be protected through our legislative and judicial systems, whether one is rich or poor, has been an essential part of the foundation for building the United States as nation. This has become a model to many other nations in which secure land tenure and property rights are still a distant goal, especially for the economically underprivileged citizens of those nations. Secure private property rights are especially important to responsible stewardship of forest lands, where the benefits of investments in sound management are long in coming and the impacts of poor management are long-lasting. In developing countries, the establishment and the fair, reliable enforcement of private property rights may be the single most important stepping-stone toward sustainable forest management.

But 200 years after the signing of the Constitution, there is still a lively debate over how best to protect the property rights of private forest owners while still protecting the important public values that these forests also provide. Perhaps more than any other category of land use, forests play an essential role in protecting values that benefit not only the owners, but also the public at large—water supply, wildlife habitat, and flood control, to name just a few. Even the resources that are of direct commercial value to the owner represent important public values, especially given the large proportion of U.S. forests in private ownership. For example, the vast majority of the wood supply in the U.S. South, one of the most productive timber-growing regions in the world, comes from private forest lands. Timber supply from private lands helps to maintain moderate prices for wood products domestically, contributing to such broad social goals as affordable housing. But forest products are also the third most valuable U.S. export, with broad public benefits in terms of lowering the international balance of payments and contributing to a stronger U.S. economy.

In spite of all this, private forest land in many areas of the country is being lost to development at a record rate. In many other areas, it is continually "on the edge" and vulnerable to conversion to nonforest land uses. Ironically, these changes are often not the first choice of the landowner, who can be forced into the decision by unforeseen circumstances or the overwhelming economics of development versus land conservation. Forest management issues notwithstanding, the fundamental policy question is how to "keep forests as forests." What are the factors responsible for forcing private forest landowners into decisions that result in the loss of forests, and what can be done about them?

For many private forest owners, the issue is simple economics—the direct financial return from their forest is less than the cost of owning the land and responsibly managing the forest. Property taxes must be paid annually, and even well-managed forests do not always produce income annually or do not produce enough to pay the taxes. Timber values increase over time, but it may be decades before the timber reaches maturity, and property taxes that include timber values have been known to force premature harvesting or overharvesting. Most localities now separate timber value from land value for the purpose of annual property taxes, and they instead tax the timber only at the time of harvest. This "severance tax" or "yield tax" approach has significantly reduced financial pressures on private forest owners.

Government regulation is another part of the financial issue, in that it can limit the income that can be generated to offset taxes and management costs. States are required to protect water quality, and such protection often requires stream buffers where timber harvesting must be reduced or avoided. Under Federal law, habitat for threatened or endangered species must be protected, and this too can lead to reductions or prohibitions of timber harvesting on some of the land, with a resulting reduction in income. Steps have been taken to minimize these kinds of effects on private forest owners, but the financial impacts of protecting these important public values on private forest lands can still be significant for some owners.

Also of major importance in the conversion of private forest lands are "death taxes"—Federal estate taxes and state inheritance taxes. These taxes affect everyone, but their implications for forest landowners often run counter to the public goal of land conservation on private forests. To pay these taxes within one year of the owner's death, heirs are often forced to sell all or part of the tract or immediately liquidate a significant portion of the timber. Recent tax reforms aimed at reducing the Federal estate tax burden by increasing the threshold of estate value that is exempt from taxes has been of only limited value to forest landowners because the value of the land and timber together often significantly exceeds the threshold. Further efforts to address this issue have been caught up in the broader issues surrounding Federal estate tax reform. The issue is unlikely to be dealt with effectively without reform that is targeted specifically to the needs of family forests, in light of the broad public benefits of better enabling landowners to continue conserving these lands as forests.

Conservation easements have become perhaps the single

most widely used mechanism for addressing the effect of taxation—both property taxes and death taxes—on private forest land conservation. If the owner donates or sells all development rights on the property, the appraised value of the property usually is substantially reduced, meaning the taxes on the property will also be greatly reduced. Unless prohibited by the terms of the easement, forest management can usually continue to produce periodic income for the landowner. Conservation easements on forest lands should be considered as part of a comprehensive estate plan aimed at keeping the ownership intact for future generations, and harvesting timber in a way that is consistent with a long-term forest management plan.

Although there are many ways to improve the economics of private forest land ownership, a large, one-time cash offer from a real estate developer can seem very appealing to forest landowners and especially to heirs who may no longer live nearby or who simply have no interest in owning and managing forest land. Forest owners who are committed to the conservation of their lands as forest for the long term, again, may want to consider a conservation easement. In many instances, development rights can be sold for a substantial sum, which can then be set aside to provide for heirs. Changes in Federal tax policy to authorize conservation bonds on which the proceeds would be exempt from Federal income tax, mentioned previously, would give land trusts and other appropriate land conservation organizations a new way to raise the private capital needed to purchase development rights and conservation easements to "keep forests as forests."

Forest Management Certification

Forest certification is a voluntary, nongovernmental process by which forest products are physically identified as having come from forests that are responsibly managed. This determination is made by an independent third party (that is, neither the forest management firm itself nor its direct customer or client) based on a field evaluation of forest management practices relative to a specific set of standards.

Forest certification differs from government regulation of forest practices (for example, state forest practices laws) in that it is market-based. It allows consumers of paper and other wood products to play a part in encouraging responsible forest management by providing them a way to distinguish certified wood products from other brands in their purchase decisions. If there is strong enough consumer preference for certified wood products, the companies that manage their forests responsibly will outcompete those that do not, leaving the latter companies either to decline or to improve the quality of their forest management so that their products, too, can be marketed as "certified."

In the United States, Federal and state policies to date encourage nongovernmental organizations in the development of certification programs, but they generally keep government agencies out of the processes of setting forest practices standards or evaluating forest management in the field. Several other countries, particularly those in Europe, have established policies to prefer the import of certified wood and wood products over those not certified. Concerned that this could be interpreted as a nontariff trade barrier under the rules of the World Trade Organization (WTO), the United States has avoided establishing a policy that could be seen as limiting the importation of noncertified wood products. The de facto policy is thus to leave this to individual decisions by the companies that import wood or wood products, based on the purchasing preferences of their own customers. Nevertheless, several public forest management agencies in the United States have sought certification of their forest practices to ensure access to markets for certified wood, to provide independent assurance to citizens that public forest lands are being responsibly managed, and to educate and encourage private forest owners to voluntarily improve forest management practices on the basis of market incentives rather than government regulation.

Background

EPA defines certification as an environmental label or "ecolabel" used to communicate information to consumers (Environmental Protection Agency 1993). Certification is an assurance that an environmental claim about a product or management system meets specified criteria. This usually involves an objective assessment of a forest management operation by an independent certifier. Some form of mark, label, or stamp attached directly to the product is then used to inform consumers of a passing grade or endorsement. The objective of certification is to link the informed consumer with products produced in an environmentally and socially responsible manner (Cabarle et al. 1995).

Independent, third-party certification did not begin with forests. In recent years, environmental considerations have been incorporated into the production of many kinds of consumer goods, and producers' claims that their products are "environmentally safe," "ozone-friendly," "dolphin-safe," "recyclable," or "made from recycled materials" have been used to market these products to the growing number of consumers who, given the choice, select products considered less damaging to the natural environment. The Federal Trade Commission regulates advertising practices regarding such "green" claims (Rathe 1992). Since neither the consumer nor the retailer is in a position to know which claims are valid and which are not, retailers often turn to independent laboratories or other organizations, such as Scientific Certification Systems, that investigate claims on everything from aerosols to pesticides to canned tuna.

Forest management certification is just one aspect of this effort to accomplish the following objectives:

- Offer consumers the opportunity to purchase from producers that have made a concerted effort to minimize impacts on the natural environment.
- Provide independent assurance that producers' claims are based on more than just empty promises.

The initial impetus for "green certification" in forestry arose from the boycott of imported tropical timbers by European consumers in the 1980s. The boycott was intended to decrease demand for this timber and reduce tropical deforestation. However, this "blunt-instrument" approach made no distinction between tropical timber obtained through exploitive means and timber harvested through responsible forest management. It equally punished companies attempting to practice renewable resource management and companies that had never given the issue a second thought, with unnecessary and unwarranted economic impacts on struggling enterprises in some of the world's poorest developing countries. Recognizing this, representatives of European environmental organizations met with forest industry representatives and tropical timber exporters to examine conditions under which tropical timbers could be produced and deemed acceptable for importation into Europe.

In the 1990s, two other international events contributed directly to the development of forest management certification as we know it today: the 1992 United Nations Conference on Environment and Development (UNCED) in Rio de Janeiro and the Uruguay Round of the General Agreement on Trade and Tariffs

(GATT). At UNCED (also called the Earth Summit), the world's governments came to consensus on the goal of sustainable development, acknowledging that sound environments and economies are inextricably linked. This included international acceptance of the "Forest Principles" section of the agreement as well as a chapter of the UNCED document, which is called *Agenda 21*, entitled "Combating Deforestation." Both embraced the concept of sustainable management of the world's forests to meet current needs without compromising the ability of current and future generations to meet their own needs (Society of American Foresters 1995). For a time, Germany, the Netherlands, and the United Kingdom considered trade restrictions in the form of requiring certification of timber originating from tropical countries. This was opposed by tropical timber-exporting nations, which asserted that any such rules should apply to all internationally traded timber regardless of source. In addition, GATT prohibits the use of trade restrictions based on methods of production to discriminate among "like products," thus preventing the use of any government-imposed tariffs, bans, or quotas in favor of sustainably produced wood.

In the absence of governmental action, the World Wide Fund for Nature brought together representatives of forest industry and environmental organizations in 1993 to form the Forest Stewardship Council (FSC), whose purpose is to "support environmentally appropriate, socially beneficial, and economically viable management of the world's forests." The FSC adopted a set of Principles and Criteria to apply to the management of tropical, temperate, and boreal forests worldwide, and it established a process for on-the-ground "performance-based" third-party verification of a forestry operation's adherence to these principles and criteria. Following successful third-party assessment, the producer is entitled to affix the FSC label directly to the product to inform consumers that the product came from a forest managed in accordance with FSC principles and criteria.

Also in 1993, the International Standards Organization (ISO) created a technical committee to develop standards and guidelines for sustainable forest management, which was carried out by the Canadian Standards Association (Hoberg 1999). ISO standards specify what processes and procedures a company needs to have in place to produce a quality product, but they do not certify actual performance under these procedures or the quality of the product itself. "Rather, the ISO certification demonstrates that the company has adopted quality management processes that are consistent and repeatable systems certification" (Berg and Ol-szewski 1995). The ISO 14001 standards are intended to document that a process or system ensuring continuous improvement in forest management exists and that management is committed to environmental performance and the achievement of sustainable forestry over time.

Meanwhile, the leading forest industry trade association in the United States, the American Forest and Paper Association (AF&PA) received the results of a nationwide survey of public perceptions of the forest products industry, which concluded that Americans were looking for tangible evidence of substantive improvement in the way the industry managed its forests. AF&PA responded with the Sustainable Forestry Initiative (SFI), a set of basic principles of responsible forest management that, by 1996, all member companies had committed to adopt as a condition for continued membership in and representation by AF&PA. Verification of actual performance was first-party, meaning it was based on the assurances by a company itself that it was in fact managing in accordance with SFI principles.

A diverse Expert Review Panel was established to advise the AF&PA board of directors on further improvements to SFI, and by 1999 the panel had persuaded the association to make several important changes. First, the SFI principles were converted to a formal standard consistent with ISO 14001. Second, AF&PA developed a licensing program whereby nonmember companies can participate in the SFI standard. Third, SFI program participants could verify their compliance with the SFI standard according to generally accepted auditing and verification procedures similar to the ISO 14001 auditing standards (Berg and Cantrell 1999). In 2001, a Sustainable Forestry Board was established to take the place of the Expert Review Panel, providing a greater degree of independence in improving the features of the SFI program. Other changes continue to be made as this and other certification programs continue to evolve and adapt to changing markets and changing perceptions of what constitutes responsible forest management.

Recent Policy Developments

Forest management certification in the United States is in a period of sorting out among several different systems, each with somewhat different objectives, costs, and benefits. U.S. forest landowners—corporations, private forest landowners, Federal and state agencies, and tribes—are evaluating the various systems to determine which systems are most consistent with their own needs and objectives.

Comparisons of the two major independent, third-party certification systems in the United States (FSC and SFI) indicate that there continue to be important differences between them, some of which are intentional. For example, FSC focuses on rewarding exemplary forest management, places greater emphasis on ecological and social components, and generally has more credibility with environmental organizations and the general public. On the other hand, SFI seeks to bring along virtually every forest products enterprise in the United States, steadily improving forest management practices by all AF&PA member companies and their suppliers (Mater, Price, and Sample 2002; Meridian Institute 2001).

Competitive as they are, the two systems are also gradually converging on certain key features. SFI has quickly evolved from being simply a code of conduct that AF&PA members pledge to abide by; much like FSC, it now requires independent, third-party verification of a company's adherence to specific standards for use of the SFI label on its products. FSC is gradually moving toward the continuous-improvement approach to the standards by which it evaluates forest enterprises, long a characteristic of SFI, and is striving to establish a more stable, inclusive approach to modifying its standards over time.

Both systems still face similar and substantial challenges to achieving their long-term objectives. Both were aimed initially at relatively large forest enterprises, and neither has been able to successfully adapt itself to the needs of private forest landowners, who generally have a limited ability to pay the costs of certification but who collectively own 60 percent of the forest land in the United States. Primary and secondary wood products manufacturers will continue to have a difficult time becoming certified if a significant fraction of their raw material comes from noncertified forest enterprises. Neither system has created the kind of awareness among American consumers that FSC has gained among many European consumers. Although several major wood products retailers have endorsed certification, particularly retailers that market their products internationally, it is ultimately the preference of consumers for certified wood products that will determine the success of this market-based mechanism.

The incentives for public sector forestry organizations are somewhat different. Several Native American tribal governments in the United States have sought to have their forest management activities certified as a way of gaining greater independence from Federal authorities. Many tribes in the United States own substantial areas of forest land, but the areas are often managed on the tribes' behalf by the Bureau of Indian Affairs, an agency in the U.S. Department of the Interior. In 1990, Congress passed the National Indian Forest Resource Management Act (NIFRMA) to provide the tribes greater autonomy in the management of their forests, but the act also required an independent evaluation every ten years to assess forest conditions (25 USC 3101–3120). The second Indian Forest Management Assessment, conducted in 2002, used a comprehensive set of field data gathered by teams of forestry experts whose primary purpose was to evaluate the tribes for possible certification. For some tribes, the process led to successful certification. For others, the process provided valuable guidance on how best to improve their forest practices and continue along the path toward greater autonomy and self-reliance.

State forestry agencies that have sought certification have often realized unexpected benefits in addition to broadening their markets to include those for certified wood. State agencies, like all public forestry agencies in recent years, have come under increased scrutiny and criticism for their forest practices and skepticism regarding their claims of sound, responsible forest management. Through independent third-party review and evaluation of their forestry activities, the agencies have been helped to better address the concerns of even the most skeptical critics and to provide an extra measure of assurance to the other citizens to which the agency is responsible. State forest lands, when certified, also become extensive demonstration forests used to educate private forest owners about responsible forest practices. In states with forest practices laws administered by these agencies, certification of private lands often can relieve the state forestry agencies of a significant burden for monitoring forestry operations on these lands. Certification requirements often meet or exceed those of the state forest practices law, meaning that certification provides prima facie proof that the private forest owner is also complying with the law.

Federal lands remain without question the most controversial when it comes to forest certification. Federal land management agencies like the Forest Service and Bureau of Land Management have generally not sought independent, third-party certification, having been discouraged from doing so by both the forest products industry and major environmental organizations. Forest management standards and guidelines are established in the process of developing a land management plan for each national forest every 10 to 15 years under the terms of the National Forest Management Act. It is possible in the future that national forests may seek independent verification of their adherence to these standards and guidelines, thereby attaining many of the same benefits that state forestry agencies have found without becoming entangled in the competitive and maturing FSC and SFI programs, which are still oriented primarily toward private forestry enterprises.

Forests and Water

The protection of forested watersheds was one of the first motivators for improving forest management in the United States. The Organic Administration Act of 1897, which defined the purposes for which National Forest lands were to be established and managed, mentions first and foremost "to improve and protect the forest within and boundaries" and second "for the purpose of securing favorable conditions of water flows" (16 USC 475). A key concern at the time was the protection of watersheds of navigable waterways to maintain commerce. A second major concern was the control of flooding from cut-over watersheds, which had been responsible for several damaging floods.

Watershed protection and the restoration of damaged watersheds are poised to become perhaps the most important issues in forest management in the twenty-first century. Forests help manage the timing and flow of water supplies for agricultural, industrial, and municipal purposes, and they protect water quality for human consumption. It is becoming particularly challenging to manage forest watersheds to maintain fish habitat, especially for threatened and endangered species. With continuing population growth and the influence of development and pollution, forests will become increasingly valuable for their ability to protect watersheds—not only in arid regions of the western United States where water is in extremely short supply relative to human needs, but also in the areas surrounding cities and towns in the eastern United States.

Background

Forests have remarkable capabilities to retain, release, and purify water. Forest soils absorb and retain water from rain and snow and slowly release that water into streams and lakes over time. Forested watersheds are also massive filters that produce high-quality water in terms of clarity and chemical content. In more mountainous regions, forests can have a tremendous influence on the timing and quantity of spring snowmelt for agricultural and municipal uses. Given the close relationship between forests and water, it is no wonder that many of the key U.S. forest policies relate directly to water.

The relationship between forests and water takes two forms: water quantity and water quality. Water quantity refers to the amount of water produced by forests and by manipulating forest cover. Water quality refers to the amount of particulates, nutrients, and pollutants in water bodies (for example, streams, rivers, lakes, and wetlands) caused by production activities like building forest roads and harvesting timber. Both water quantity and quality continue to be dominant concerns in U.S. forestry.

Forest watershed protection was one of the first motivators for improving forest management in the United States. Because forests regulate the timing and release of water into streams and rivers, forest conservation was deemed necessary for maintaining navigable waterways as a primary source of commercial transport through the early 1900s. Not only do exploited forests contribute to flooding, they produce large amounts of sediment that can render waterways unnavigable, silt up agricultural irrigation canals, and clog intake pipes for municipal water supplies. For these reasons, forest watershed protection was the highest priority defined by the Organic Administration Act of 1897, which established the direction for managing the National Forest System. Virtually all rivers in the western United States and many in the eastern United States have their source on National Forests. Watershed protection and management remains a high priority for the Forest Service today.

Population and industrial growth throughout the twentieth century created greater demands for water. Forest managers and watershed scientists experimented with different forest-cutting techniques to manage the timing and flow of water. The first tests of the effect of forest-cutting patterns on water flow were conducted in experimental forests managed by the Forest Service, such as Hubbard Brook in New Hampshire and Fraser in Colorado. In general, water yield can be increased over the short term by as much as 10 percent by creating small, scattered openings in the forest. However, many factors can change this increase, and serious questions remain whether this increase is sustainable over the long term and over a large area. It should be noted that the cutting experiments took place in fairly small areas.

The emergence of forest water quality as a policy issue was tied to general concerns over water pollution throughout the 1960s and early 1970s. The Federal Water Pollution Control Act of 1948 primarily addressed point sources of pollution such as industrial drainpipes. Scientists and policymakers soon recognized that nonpoint sources contribute more significantly to water pollution. Nonpoint sources generate pollutants attributable to many dispersed activities that cannot be precisely identified. Streets, parking lots, lawns, and agricultural lands are considered nonpoint sources. Forestry activities such as road building, road use, and timber harvesting are also considered nonpoint sources. The Clean Water Act (CWA) of 1972 and the CWA Amendments of 1987 directly addressed nonpoint sources of water pollution and prompted states to look at forestry practices more closely. Primary pollutants from forestry activities are sediment, nutrients (for example, nitrogen and phosphorus), petroleum from machinery, and chemical pesticides and herbicides. In response, several states enacted forest practices acts to regulate private forestry activities. Many other states work cooperatively with forest industry and private landowner associations to develop and implement voluntary best management practices, or BMPs, to minimize forestry-related sources of water pollution.

Recent Policy Developments

In connection with water quantity, one of the most controversial policy developments is the assertion of Federal reserve "instream flow" rights. This controversy has emerged in the western United States, where water law is governed by a Prior-Appropriations Doctrine that reflects a "first in time, first in right" allocation of who gets water. The Federal government is a relative latecomer to the water rights system, with the prime water rights having been appropriated by early miners, farmers, and other homesteaders and industries. Hence, the Federal government is considered under the prior-appropriations doctrine to hold junior water rights. One of the conditions of holding water rights is that the water has to be used—"use it or lose it." Water use typically means diverting water out of the river to be applied to some use before it is returned to the river. However, the Forest Service has recently asserted a claim to watercourses running through National Forest lands, requiring that a certain amount of water be left in streams and rivers to support fish and recreation, among other values. This claim, in allowing a junior water rights holder to usurp senior rights holders and in leaving water in streams and rivers rather than diverting it for use, runs counter in both respects to the Prior-Appropriations Doctrine. Although several specific disputes have been resolved out of court, serious legal questions remain that need to be resolved across the West.

In terms of water quality, both states and Federal land management agencies have paid special attention to forestry activities in areas adjacent to streams and rivers. Forest stands along these riparian areas serve multiple watershed functions: they stabilize the bank, provide shade, and filter runoff after storms. They also support a broad range of habitat and wildlife corridors that contribute to biological diversity. The Northwest Forest Plan (1994), which sets strategic priorities for the management of Federal lands west of the Cascade Mountains in the Pacific Northwest and in northwest California, has some of the most restrictive riparian forest protections in the world, protecting 300 feet on each side of many streams and rivers from any kind of forestry activities. Although such restrictions have not been placed on privately owned riparian areas, the issue remains at the front of both policymakers' and landowners' agendas. Even in traditionally laissez-faire states like Virginia, concern about forestry activities in riparian areas has prompted more restrictive policies on private lands.

One ongoing political battle has been over the designation of forestry as a point source of water pollution. The Environmental Protection Agency has on several occasions sought to put forestry under its purview. Doing so would subject forestry on both public and private lands to permits and regulations to reduce the "total maximum daily load" (TMDL) of pollutants in any given water body. In 2002, the EPA once again proposed to place forestry under the TMDL requirements. Compared with agriculture and urban development—both classified as nonpoint sources of water pollution—forestry contributes very little in terms of chemical, nutrient, and sediment pollution to streams, rivers, and lakes. Herbicides and pesticides are typically applied on young forests over the first five years or so, but they are not used after the young trees have reached a certain age and height. In contrast, agricultural crops often require multiple applications of pesticides and herbicides in a single year, year after year. Forest road building, harvesting, and processing are episodic, typically lasting several weeks before operations end. Federal and state forest practice regulations restrict these activities and often go so far as to require site rehabilitation after the operations end—that is, road obliteration and replanting. Most of these regulations are in place to directly address water quality issues stemming from forestry. In audit after audit of forest practices on public and private land, compliance with best management practices is typically in the 90 percent range. In the case of urban development, land disturbance activities are typically permanent and contribute to water quality problems on an ongoing basis.

Both water quantity and water quality have become issues after the catastrophic wildfire seasons in 2000 and 2002, especially in the western United States but also in Florida and other parts of the Southeast. In Colorado, which is the headwaters for the Colorado, Platte, Arkansas, and Rio Grande Rivers, large wildfires pose extreme threats to water quantity and water quality, because scorched forest lands lose their capacity to retain and release water during and after storms. In 1996 in Colorado, massive sedimentation and debris flow resulted when heavy rain fell in the area of the Buffalo Creek fire, clogging the intake valves in the Strontia Springs Reservoir, which serves as a municipal water source for the Denver metropolitan area. Repairs reached into the millions of dollars. The ability of severely burned watersheds to retain and release snowmelt into streams and rivers is hampered, compounding the effect of drought and increasing human demand on diminishing water supply.

Conclusion

Water quantity and quality will continue to be high-priority issues in U.S. forest policy, especially as population growth and development increase pressure to protect and augment existing water sources. In fact, water is poised to become perhaps the most important issue in forest management in the twenty-first century. The relationship between forests and water is compounded by the societal demand to protect biological diversity, manage catastrophic wildland forest fire risk, and increase the human presence in previously undeveloped forest land. An ultimate future challenge will be to optimize forest management to capture, store, and release water that is in extremely short supply relative to human needs.

Forests and Atmospheric Carbon

Forests provide the world with more than wood; they abound with a broad array of environmental services, many of which are essential to the survival of all life on Earth. Yet, many of these services have no price and so are not accounted for in the market or even in public policy. One of these essential services is the storing and processing of atmospheric carbon, which is produced by the cumbustion of fossil fuels and is one of the leading contributors to global climate change. Because the value of forests to process and store carbon is rarely factored into the financial side of forest management decision-making, the value of a forest's services to society is not fully accounted for.

Recent initiatives such as the Kyoto Protocol seek to change this (UNFCCC 1997). One area of focus is on developing an accurate, reliable, transferable accounting system to calculate the full value of forest environmental services so that the financial side of forest management decision-making is more fully informed and forest conservation efforts are appropriately rewarded. When a financial price is assigned to a forested area's services, those services can be bought, sold, and traded like any other service. A second focal point is on developing the institutional mechanisms to bring forest environmental services into the market exchange, such as trading carbon credits among emitters of atmospheric carbon, such as coal-fired electricity generating plants. By purchasing a certain number of carbon credits based on its annual carbon emissions, the emitter effectively pays for forest lands that process and store the corresponding quantity of annual carbon emissions. As unconventional as this may seem, the calculation, designation, and exchange of forest environmental services on the open market is being seriously considered around the world.

Background

Forests are one of Earth's great respirators, converting carbon in the atmosphere into life-giving oxygen. Second only to oceans, the world's forests process more carbon than any other ecological system. Forests not only convert atmospheric carbon to oxygen, they also store an enormous amount of carbon that would otherwise be released into the atmosphere—more than half of the world's terrestrial carbon is stored in forests. Forests store carbon not only in the leaves, branches, and stems of trees, but in the soil as well. There are three categories of activities through which forest conservation can help reduce atmospheric carbon (Bass et al. 2000):

1. Carbon sequestration
   - Afforestation, reforestation, and restoration of degraded lands.
   - Improved silvicultural techniques to increase growth rates.
   - Implementation of forestry practices on agricultural lands to boost those lands' ability to sequester carbon.
2. Carbon conservation
   - Conservation of biomass and soil carbon in existing forests (that is, by forestalling forest cutting).
   - Improved harvesting practices to reduce effects on remaining biomass and soils.
   - Improved efficiency of wood processing to reduce waste.
   - Fire protection and more effective use of burning in both forest and agricultural systems.
3. Carbon substitution
   - Increased conversion of forest biomass into durable wood products for use in place of energy-intensive materials such as concrete or aluminum in housing construction.
   - Increased use of biofuels to generate power.
   - Enhanced use of harvesting waste as feedstock.

Each category varies in how effective it is and over what period of time. Carbon sequestration through the conservation of biomass and soil carbon in *existing* forests has the largest and most immediate positive effect on reducing atmospheric carbon. Carbon sequestration through reforestation or afforestation takes place over a much longer period of time. Substituting biofuels for fossil fuels slows, but does not reverse, carbon emissions. Conservation of existing forests allows mature forests to continue sequestering carbon at a high rate and also helps avoid major carbon emissions associated with deforestation. Since the early 1980s, as much as one-quarter of all carbon emissions from human activities has come from deforestation and the burning of forests, primarily in the tropics.

In the United States, economic pressures continue to result in the conversion of significant areas of forest land to housing developments and other nonforest land uses. Even considering all the National Forests and forest industry lands managed with a long-term perspective, half of all U.S. forests are in small private ownerships (Smith et al. 2001). Factors such as property taxes often make it difficult for these owners to achieve a positive financial return on their lands, and estate and inheritance taxes can make it nearly impossible for heirs to maintain these forest lands unless they fragment them or convert them to nonforest uses. With the loss of these forest lands, there is also the loss of the numerous ecosystem services these forests provide, such as watershed protection, wildlife habitat, and carbon sequestration. These are important public values that private forests routinely provide. When private forests disappear, the public values disappear too.

To help ensure that private forest ecosystems continue to provide these public values, new mechanisms are being developed to better recognize the values and to translate them into financial values (Daily 1997). Some of these are mechanisms for direct compensation of private forest landowners. For example, in 1989, the EPA informed New York City that its drinking water quality was declining to the point of requiring construction of sophisticated water treatment plants at an estimated cost of $6 billion to $8 billion (Daily and Ellison 2002). Faced with so costly a project, the city government began exploring alternatives and decided to embark on an aggressive land conservation program to protect 2,000 square miles of upstate forest land in the Hudson River watershed through acquisition of land and conservation easements, as well as stricter controls on new development. The land conservation program was essentially a transfer of about $1.5 billion from water users in New York City to upstate communities and landowners. New York City will clearly want to rely on the water filtration services of its forested watershed for as long as possible before it switches to the technological solution. Ecosystem services in the New York City watershed are being better protected today, not only because we have a better estimate of their value, but because some of that value is actually being captured by the private forest owners whose lands provide watershed protection services.

With deepening concern over global climate change and recognition of the important role of forests in mitigating atmospheric carbon dioxide, there is a potential that this value, too, will be recognized in the marketplace in ways that will promote improved forest conservation. The potential role of forest conservation and management to decrease greenhouse gases in the atmosphere suggests that forests will eventually have to be part of any effective strategy to mitigate climate change. Government policies and land use regulation may play a part, but it is possible that the greatest influence may be private, voluntary participation in innovative capital markets for ecosystem services that are just now being developed.

Recent Policy Developments

The United Nations Framework Convention on Climate Change (FCCC) of 1992 sets broad goals for reducing greenhouse gas emissions and enhancing carbon "sinks"—ecosystems that can process and store large quantities of atmospheric carbon, like forests (UNFCCC 1992). The FCCC of 1992 was signed by virtually every country around the world, including the United States. In December 1997, delegates from 160 nations met in Kyoto to continue work on the FCCC, resulting in the objective of reducing greenhouse gas emissions to 5 percent or more below 1990 levels by 2012. Although there was no formal agreement on how to account for forests as carbon sinks, it is clear that they will play a key role in any global strategy to mitigate greenhouse gases. Aside from storing about half of the world's terrestrial carbon, they account for about 80 percent of carbon exchange between terrestrial ecosystems and the atmosphere, absorbing some 3 billion tons of carbon a year. Aside from oceans, few carbon sinks can match the carbon processing and storage capacity of forests.

Nongovernmental organizations have generally taken the lead on linking forest conservation to carbon reduction. Organizations involved are typically industries seeking to offset their carbon emissions, intermediary nongovernmental organizations (NGOs), and specific forest managers. Some example projects include:

> **Rio Bravo Conservation and Management Area, Belize**
> *Objective:* Protection of 14,000 hectares of endangered
> forest and development of a sustainable
> management program for another 46,000 hectares;
> all of this land was slated to be cleared for
> agriculture.
> *Forest area:* 60,000 hectares (148,000 acres).
> *Estimated carbon credit and price:* 2.5 million tons @
> US$3/ton (US$7.5 million).

Baseline assumed: Clearing of entire area within five years.
Partners: Programme for Belize (Belizean NGO) and group of U.S. utilities.

Scolel Te Community Forestry Project, Mexico

Objective: Develop systems by which small farmers can gain access to carbon markets (activities include reforestation, forest restoration, and agroforestry).
Forest area: 300 hectares (741 acres).
Estimated carbon credit and price: 312 tons @ US$12/ton (US$3,744).
Baseline assumed: Mean carbon storage under continuation of existing land use.
Partners: Ambio (Mexican NGO), El Colegio de la Frontera, University of Edinburgh, and UK Department for International Development.

Carbon Sink Project in Matto Grosso, Brazil

Objective: Sequester carbon in a plantation of native species established on degraded land.
Forest area: 5,000 hectares (12,300 acres).
Estimated carbon credit and price: "Maximize sequestration"; US$10 million.
Baseline assumed: Continuation of previous land use.
Partners: Peugot, Instituto pro Natura (Brazilian NGO), Office National des Forêts of France.

Sustainable Management of Reserva Forestal Malleco, Chile

Objective: Promote sustainable management of this state-owned forest and prove the feasibility of the new management regime both in economic and carbon sequestration terms.
Forest area: 16,625 hectares (41,000 acres).
Estimated carbon credit and price: To be determined.
Baseline assumed: Continuation of previous land use.
Partners: French Fund for the Environment, Corporación Nacional Forestal of Chile, and Office National des Forêts of France.

In each of these examples, a utility or industry seeking to mitigate its carbon emissions has sought out and then directly implemented a major project to sequester the corresponding amount of carbon it expects to emit. One drawback of a project-by-project approach is that the twin problems of carbon emissions and degradation or loss of forest ecosystems are occurring at a global scale. Intermediary nongovernmental organizations are just not well positioned to facilitate each and every possible industry-forest exchange.

In response to the Kyoto standards set in 1997, many countries are exploring creating a carbon exchange system that would operate much like any other commodities futures market. The international carbon credit trading system now being developed by Australia, Canada, Mexico, and the European Union is based on a highly successful sulfur dioxide (SO_2) "cap-and-trade" system developed by the United States and aimed at reducing the causes of acid rain. The SO_2 system was established by the Clean Air Act Amendments of 1990 under President George H. W. Bush and is administered by the EPA. Under these amendments, EPA issued permits to utilities for a certain level of SO_2 emissions and issued the right to trade these permits with other utilities. It was more expensive for older-technology power plants to achieve the required reductions than it was for newer, more efficient power plants. This stimulated an active trading of SO_2 credits from new plants to old, but it also prompted the new plants to become even more efficient so as to have more unused permits to sell. Ten years later, the market in SO_2 permits had grown to $3 billion, overall SO_2 emissions by the utility industry as a whole were significantly *lower* than EPA targets, and the reduction had been accomplished at about one-tenth of the predicted cost.

The 2001 decision by President George W. Bush to withdraw the United States from participation in the Kyoto Protocol has significantly slowed progress toward developing a corresponding cap-and-trade system for carbon dioxide (CO_2) emissions in the United States and internationally. The United States is the source of about one-quarter of the world's carbon emissions; without U.S. participation, no such system is likely to be effective in reducing global atmospheric carbon.

This has not stopped individual states within the United States from acting, however. More than half the states have adopted voluntary or mandatory programs for reducing carbon emissions. In February, seven state attorneys general announced that they would sue the EPA for failing to enforce the Clean Air Act, and they called for new Federal regulations on emissions of CO_2 and other greenhouse gases (Lee 2003). In 2003, Senators

John McCain (R-Arizona) and Joseph Lieberman (D-Connecticut) announced proposed legislation (S.139) that would establish a national cap-and-trade system for carbon, modeled closely on the successful system developed for reducing SO_2 emissions (Pianin 2003).

Also in 2003, the Chicago Climate Exchange (CCX) opened for business and became the first financial market in the United States to begin trading carbon credits like any other commodity (Behr and Pianin 2003). The goal of the CCX is to implement a voluntary, private cap-and-trade pilot program for reducing and trading greenhouse gas emissions. CCX will administer this program for carbon-emitting industries and for farm and forest carbon sinks primarily in North America. To foster international emissions trading, forestry operations in Brazil can also participate. CCX's founding member companies have made a voluntary binding commitment to reduce their greenhouse gas emissions by 4 percent by 2006 (MeadWestvaco 2002).

Among CCX's 14 founding member organizations are 4 major forest products corporations (International Paper, Mead-Westvaco, Temple-Inland, and Stora Enso), 3 of them based in the United States (Chicago Climate Exchange 2003). There is more work to do at the national policy level in order to create the kind of framework that will be required to make this program work in the long run. But clearly, according to CCX, there are market players positioned to "make investments that will spur innovation, raise productivity, and lead to new profit centers." And the founding members of CCX have a unique opportunity, through this bold and innovative pilot program, to help shape the future policy framework within which their organizations will operate.

Conclusion

Forests provide essential environmental services that society often takes for granted. Keeping forests growing in natural conditions ensures the sustainability of these services. But based on current market values of timber alone, many areas of forest land in the United States and around the world are still not overly valuable. Many private landowners continue to have great difficulty just breaking even on their forestry investments. Such losses often lead to the conversion of forest land to nonforest uses such as agriculture or development. Public forest managers are having to do more on steadily declining budgets. Development of financial markets for the environmental services provided by forests in processing and storing carbon has the potential for increasing the values derived from forests and thus improving forest conservation in the United States and around the world.

Maintaining and enhancing forests that are not under strict production objectives remains one of the most important challenges for the future. Wood fiber production is likely to increase on intensively managed industrial plantations. Because of the economies of scale involved in industrial wood production, most nonindustrial and public forest landowners probably will not be able to compete in either the wood products market or the financial investment market solely based on wood production values.

Forest Biotechnology

Biotechnology is a relatively new issue area in forestry—one in which the policy framework is still being formed. What little policy is currently in place is largely an offshoot of policies originally developed to guide the use of agricultural biotechnology. Forest biotechnology has significant potential benefits in terms of increased growth to meet expanding needs for wood and fiber. It also has the potential to improve environmental performance through lower energy requirements for production processes and to reduce use of chemical fertilizers and pesticides. There are also potential risks, however, that scientists are unable to fully define or predict. Public concern about these risks is likely to result in policies that will require rigorous review of every new application of genetically modified tree species in the United States.

Background

Genetic improvement of commercial tree species has long been an important aspect of forest management for wood production. In traditional genetic improvement, individual trees exhibiting certain favorable characteristics—such as straightness, frost-hardiness, or resistance to diseases common to the species—are crossed with other individuals of the species through controlled combinations of pollen from one and seed from the other. Over time, this has produced generations of offspring that grow faster, produce higher-quality wood, and are less susceptible to damage or early mortality from insect pests and disease. These offspring are propagated in forest nurseries, often by the thousands, to produce the seedlings that are planted to regenerate forest areas following timber harvests. Special varieties of natural tree species are also bred for particular purposes such as afforestation of desertified areas, reclamation of mining spoils, or planting among agricultural crops in agroforestry operations.

It is important to note that in this type of genetic improvement, the collection of genes inherent to the species of tree (its genome) remains unchanged from that of wild individuals of the same species. Even after many generations of genetic tree improvement, in which individuals exhibiting desired traits have been favored repeatedly over individuals not exhibiting those traits, the composition of genes—and the DNA that serves as the building blocks of these genes—is still the same.

The current policy issues relating to forest biotechnology revolve around a different approach to genetic modification—that of actually altering the DNA and, in effect, creating a new species. Scientists working first with agricultural crops discovered that, by clipping out certain sections of a species' DNA and replacing them with DNA of another species, they could engineer species of grains, vegetables, and other agricultural crops that grew faster and larger; survived over a wider range of soil types, climates, and moisture regimes; were of greater nutritional value; and could even protect themselves from common insects and diseases. Scientists saw the potential for a second "green revolution" in which food production could be markedly increased to meet the needs of a growing world population. Among other benefits were lower requirements for chemical fertilization and pesticide application, reducing energy requirements and agricultural effects on environmental values such as water quality and aquatic habitat.

However, concerns were raised about the unintended, unpredictable, and potentially harmful effects of such genetic engineering. Would corn designed to produce natural toxins that killed corn borers also wipe out certain species of butterflies whose larvae also feed on corn (Prasad 1999)? Could the corn itself have unanticipated negative health effects on people who ate it?

Genetic engineering is a complex and evolving science. The genome of a given plant or animal species is composed of thousands of individual genes, each of which has a particular purpose or role. Some genes express themselves in not one, but multiple ways. Other important expressions arise from clusters of certain genes, rather than from just one. The characterization or "mapping" of genomes—for species from rice to humans—is one of the great scientific challenges of the twenty-first century. For now, however, the inability of scientists to say definitively what all the effects of a given genetic modification will be—whether for the environment or for the consumer—has led to public concern, countries banning the importation of food derived from genetically modified crops, and even calls for a prohibition on any further research or field trials of genetically modified species (Bruce and Palfreyman 1998).

Potential human health effects are of minimal concern in forest biotechnology when compared with agricultural uses, but concern over its potential environmental effects may be even greater. Unlike most agricultural crops, genetically modified (GM) tree species often would be grown in close proximity to wild populations of the same tree species. This leads to the possibility of interbreeding and the escape of genetic modifications into the outside environment, with unknown consequences for the long-term survival of the original species or of other plant and animal species that live in association in natural ecosystems.

Scientists have attempted to address this issue by programming delayed seed or pollen production in GM trees (so that there is little chance for genetic interaction with wild populations) or by making the GM trees sterile altogether (Bajaj 2000). Nevertheless, the continuing scientific uncertainty over the effects of GM trees is seen by many as constituting an unacceptable risk, and public concern will continue to pressure political leaders to develop new policies to guide the development of forest biotechnology.

Recent Policy Developments

U.S. policies governing research and commercial use of genetically modified trees are largely indirect, having been developed primarily for application to agricultural biotechnology (Friedman 2002). At least three Federal agencies are currently involved in some way in regulating the use of GM crops and forest trees.

The U.S. Department of Agriculture, through its Animal and Plant Health Inspection Service (APHIS), has regulatory authority over the importation or propagation of organisms that are regarded as "plant pests"—that is, that pose a danger to U.S. agricultural and tree species (Raychaudhuri and Maramorosch 1999).

To date, there has been no formal determination by APHIS that GM trees are plant pests as defined by the Plant Protection Act (7 USC 1501 [note]). There is significant public pressure on APHIS to do so.

To the extent that tree species are genetically modified to produce food, animal feeds, pharmaceuticals, or cosmetics, the U.S. Food and Drug Administration has regulatory authority to approve or disapprove of the use of these products in the United States.

The EPA, through its authority to regulate the development and application of pesticides (7 USC 136 et seq. 1996), also asserts authority over tree species that have been genetically modified to produce their own toxins to guard against insect or disease organisms. There has also been experimentation with genetically engineering certain tree species to absorb serious soil contaminants such as lead, arsenic, and mercury, thereby providing a new tool for the cleanup of toxic wastes. This sort of "bioremediation" also falls under the regulatory jurisdiction of EPA. Most importantly, EPA has specific jurisdiction for regulating the "microbial products of biotechnology" through the Toxic Substances Control Act (15 USC 2601 et seq.). The EPA has authority to review new chemicals before they are introduced to commerce, and "intergenetic microorganisms" (microorganisms created to contain genetic material from organisms in more than one taxonomic genera) are considered new chemicals. Federal regulations now establish the manner in which EPA will review and regulate the use of intergenetic microorganisms in commerce or commercial research (62 FR 17910).

Conclusion

Thus far, there are few if any examples of large-scale commercial production of GM tree species. However, hundreds of field trials involving at least 24 species of trees are under way in more than a dozen countries around the world (Owusu 1999). International research consortia have been formed with the objective of expanding the use of GM trees to commercial scale, and at least one consortium has indicated its intent to seek approval from the U.S. Department of Agriculture and other regulatory agencies for mass plantings in the United States by 2005 (Brown 2001). Genetic modification of commercial tree species may hold valuable opportunities for supplying wood needs from a smaller area of intensively managed plantation forests, with reduced effects on water quality and other environmental values through less use of chemical fertilizers and pesticides (Arborgen 2002). Other benefits may include genetically modified forms of important tree species that are immune to diseases that have nearly eliminated the natural species, including American elm (nearly destroyed by Dutch Elm Disease), American chestnut (virtually eliminated from American forests by chestnut blight), and flowering dogwood (in sharp decline as a result of Dogwood Anthracnose).

There are potentially significant risks that are as yet ill-defined and poorly understood even by the scientists most closely involved in the ongoing research on genetic modification of trees. The long life of forest trees, as compared with annual agricultural crops, poses a special challenge for scientists because negative traits of certain genetic modifications may not become apparent for several years into the trial of a particular GM tree. Such trials may turn out to be the most time-consuming and expensive part of future research on GM trees, but they could also be essential to demonstrating to the public that all reasonable measures have been taken to minimize the biological and ecological risks associated with artificially engineering new tree species.

As with many other public policy issues involving the introduction of new technologies, future policymaking will require a careful weighing of potential benefits and potential risks, and involving diverse segments of the public who may regard these benefits and risks somewhat differently.

Previous experience with new technologies also suggests the wisdom of moving forward carefully and taking a case-by-case approach to using new technology to accomplish near-term objectives. More traditional approaches to genetic tree improvement—that is, approaches that do not involve the introduction of genetic material from other species—still have significant unrealized potential for increasing growth rates and enhancing disease resistance. The per capita consumption rate for industrial roundwood is decreasing, and wood production from plantations worldwide is continually increasing, and so little or no increase in wood prices is projected during the next few decades (Adams 2002). All of this suggests that there is little urgency to rush this new genetic modification technology into widespread use, and that scientists have time to more fully explore the potential risks of what could ultimately be a very beneficial technological achievement.

Emerging Issues and Future Policy Directions

What are some of the major trends influencing the future direction of forests, forestry, and U.S. forest policy? There will be continued concerns over sustainable wood production, protecting habitat for endangered species, protecting remaining areas of wilderness and primeval forest, and restoring and protecting forested watersheds.

New issues will arise, some of which will require new bodies of science and new institutional frameworks for policy implementation. For example, we are only now beginning to understand the role forests and forest soils play in sequestering carbon dioxide from the atmosphere and thus helping regulate global climate. Understanding how best to optimize the contribution of forests to moderating global climate change is a first step; establishing the institutional, legal, and policy framework to facilitate the monitoring of carbon storage in forests—and establishing the market mechanisms to permit the trading of "carbon credits" from carbon-storing countries to carbon-producing countries—is yet another step.

The basic questions that have been the focus of U.S. forest policy for decades will remain. In the future, as in the past, economics will be the primary driver, with policy and law as secondary influences.

Wood Supply in the Global Context

The economic globalization that has so transformed the U.S. forest products industry during the past decade can be expected to continue at an even faster pace, particularly in terms of where the United States gets its wood and wood fiber. In decreasing order of importance, future U.S. wood supplies will come from:

- Southern Hemisphere forests, particularly forest plantations in South America
- Forest plantations on industry timberlands in the U.S. South and Pacific Northwest
- Other industry timberlands and TIMOs
- Nonindustrial private forest lands in the U.S. South
- State forest lands
- Federal forest lands

Southern Hemisphere

High growth rates (even relative to the U.S. South) and lower costs for land, labor, and capital will continue to draw timberland investments to South America and other parts of the Southern Hemisphere. Lower costs for converting wood to pulp for paper and paperboard (that is, costs related to energy, labor, and environmental restrictions) will increase the proportion of imported Southern Hemisphere pulp going to U.S. paper mills. Gradually, an increasing proportion of finishing plants for commodity fiber products (for example, white business papers and kraft paper for packaging) will also be located in the Southern Hemisphere. With the continued transnational consolidation of the paper and wood fiber industry, the term *U.S. forest products company* will lose its meaning: companies' headquarters may remain here, but their facilities and markets will be global.

Industry Timberlands in the United States

The relatively high cost of land, labor, and capital in the United States will make it increasingly difficult for forest products companies to continue owning and managing large areas of forest land here. This will lead companies to retain only those lands that can be intensively managed for uniform, high-quality fiber for high-value specialty products (for example, coated stock for consumer packaging and coated papers for magazine publishing and commercial printing). The forest products industry has sold millions of acres of forest land in the United States during the past decade, and it is expected that the industry will divest millions more acres in the years to come. Many of these lands will be acquired by the timber investment management organizations (TIMOs) that invest in land on behalf of pension funds, insurance companies, and other institutional investors more interested in appreciation of financial value and reduction of investment risk than in timber supply per se. Many other areas of industry forest land, particularly those in the "growth path" of burgeoning cities in the U.S. South and Pacific Northwest, will be developed.

Other Private Forest Lands

The nonindustrial private forests that constitute the majority of the forest land in the United States will continue to be a timber supply reserve for industry in times of increased market demand. This is especially true in the U.S. South, where forest industry lands were cut heavily in the early 1990s in the wake of the sharp reduction in timber supply from Federal lands. Replanted forests in industry lands in the South are expected to once again become the major source of timber supply in that region after 2015, but until then most of their supplies will come from nonindustrial private forest (NIPF) lands. Given that there is little if any regulation of forest practices on private lands in most states in the U.S. South, the current level of public concern over the management of these forests is certain to increase. In other regions of the United States, most of the NIPF lands will continue to be managed at low to moderate intensity for multiple values and uses. These family forests will continue to produce timber sporadically, more in response to changes in owners' circumstances (retirement, college tuition, death) than to fluctuations in timber prices.

Public Forest Lands

Considering the political uncertainty and the legal and regulatory process requirements, public lands will increasingly be the most expensive place to get timber. The large transnational forest products companies that have resulted from the past decade's mergers and industry consolidation have many other ways to obtain wood closer to their manufacturing facilities and markets. Federal forests will continue to be an important source of timber supply primarily to small and medium-sized regional firms, particularly in the Intermountain West. Much of the timber supply from Federal land will be determined primarily by forest health concerns, that is, it will come from thinning to control hazardous fuels in areas prone to wildfires.

Forest Management as a Plus in Biodiversity Conservation

With continued population growth and a finite land base, dependence on wild stocks of trees will steadily decrease, and dependence on intensive culture will increase. This has already happened with agriculture and animal husbandry. Marine fisheries and forestry are among the last major human enterprises to shift from "hunting and gathering" from natural populations wherever they might occur to intensive production in smaller areas under more controlled conditions. For forests, this means more intensive management of forest plantations on the most productive forest lands and on marginal crop and pasture lands that can be profitably afforested. Not only can this reduce timber harvesting pressures on other forest areas, but it can also help create new opportunities to establish protected areas where forests harbor significant biodiversity resources or other environmental values.

Several conservation organizations regard such approaches as the best hope for protecting forest biodiversity around the world while also meeting the world's needs for paper and building products through sustainable wood production. The World Wide Fund for Nature has called on the world's ten largest forest products companies to expand the area of forest plantations by 5 million hectares annually for the next 50 years, as part of a global strategy for biodiversity conservation. Even with the doubling of current industrial roundwood demand by 2050 predicted by the UN Food and Agriculture Organization (FAO), it is estimated that 80 percent of the world's wood needs could be met from less than 20 percent of the world's forests. With wood production thus concentrated, an estimated 20 percent of the world's forests could be devoted to preserving habitat for rare, localized, threatened, and endangered species. The remaining 60 percent would be managed at low or moderate intensity for wood products and other uses.

This approach is also appealing from a broader environmental perspective, in that meeting building material needs as much as possible from a renewable, recyclable resource like wood can reduce the use of concrete, steel, and aluminum. In addition to being nonrenewable, these require far more energy to turn the raw material into usable products, exacerbating the impacts of fossil fuel burning on greenhouse gas emissions and global climate change. Greater understanding and acceptance of intensively managed forest plantations by conservationists and greater understanding and acceptance of protected forest areas by forest industry can lead to a new consensus that these are all components in a comprehensive strategy for the conservation and sustainable use of forests—and an end to the gridlock that has characterized so much of U.S. forest policy for much of the past half-century.

Putting U.S. Forest Policy in the Global Context

The world is becoming a smaller place. U.S. forest policy will become increasingly intertwined with international policymaking that takes a global perspective when assessing trends and conditions across the world's forests and facilitates local actions that meet local needs in ways consistent with moving ever closer toward the goal of sustainable use and management. In 1992, the

Earth Summit in Rio de Janeiro set the stage for national-level assessments of forest conditions and trends, with reference to Criteria and Indicators of sustainability developed in the earlier Montreal Process. Ten years later, the World Summit on Sustainable Development in Johannesburg focused on moving beyond paper resolutions to increase the demonstrable, measurable improvements in forest management at the local level. Although there is still significant room for improvement in the way U.S. forests are managed, the United States nevertheless has the potential to serve as a model of sustainable forestry that is ecologically sound, economically viable, and socially responsible. Leadership will come not only through the transfer of useful technologies and proven forest management methodologies, but through the adaptation of institutional, legal, and policy mechanisms to provide a framework for creativity and innovation. After all, it was not so long ago that the United States was still a developing country, especially in the context of forestry.

References

7 USC 1501 (note). Agricultural Risk Protection Act of 2000. 114 Stat. 358.

7 USC 136. Federal Insecticide, Fungicide and Rodenticide Act of June 25, 1947. 61 Stat. 63, as amended.

15 USC 2601 et seq. Toxic Substances Control Act of October 11, 1976. 90 Stat. 2003, as amended.

16 USC 1131 (note). Wilderness Act of September 3, 1964. 78 Stat. 890, as amended.

16 USC 1600 (note). National Forest Management Act of October 22, 1976. 90 Stat. 2949, as amended.

16 USC 1600. Forest and Rangeland Renewable Resources Management Act of August 17, 1974. 88 Stat. 476, as amended.

16 USC 528. Multiple-Use Sustained-Yield Act of June 12, 1960. 74 Stat. 215.

36 CFR 219 Planning. Subpart A. National Forest System Land and Resource Management Planning.

62 FR 17910. "Microbial Products of Biotechnology: Final Regulation under the Toxic Substances Control Act." *Federal Register,* April 11, 1997.

Adams, D. M. 2002. "Harvest, Inventory, and Stumpage Prices: Consumption Outpaces Harvest, Prices Rise Slowly." *Journal of Forestry* 100(2):26–31.

Aplet, G., N. Johnson, J. Olson, and V. A. Sample. 1993. *Defining Sustainable Forestry.* Washington, DC: Island Press.

Arborgen. 2002. "Improving Forestry through Genetics and Technology." Available at: www.arborgen.com/science.html. Accessed on April 30, 2003.

Bajaj, Y., ed. 2000. *Transgenic Trees.* Biotechnology in Agriculture and Forestry, No. 44. New York: Springer Verlag.

Banzhaf, W. 1995. "The Forest Congress: A New Way of Doing Business?" *Journal of Forestry* 93 (3): 3.

Bass, S., O. Dubois, P. Mouracosta, M. Pinard, R. Tipper, and C. Wilson. 2000. *Rural Livelihood and Carbon Management.* IIED Natural Resources Paper No. 1. London: International Institute for Economic Development.

Behr, P., and E. Pianin. 2003. "Firms Start Trading Program for Greenhouse Gas Emissions." *Washington Post,* January 17, 2003, p. A14.

Berg, S., and R. Olszewski. 1995. "Certification and Labeling: A Forest Industry Perspective." *Journal of Forestry* 93(4):30–32.

Berg, S., and T. Cantrell. 1999. "Sustainable Forestry Initiative: Toward a Higher Standard." *Journal of Forestry* 97(11):33–36 (November 1999).

Binkley, C. S. 2001. "Forestry in the Long Sweep of History." In *Sustainable Forestry and Biodiversity Conservation: Toward A New Consensus.* Washington, DC: Pinchot Institute for Conservation.

Binkley, C. S., C. F. Raper, and C. L. Washburn. 1996. "Institutional Ownership of US Timberland." *Journal of Forestry* 94(9):21–28.

Bowes, M., and J. Krutilla. 1989. *Multiple-Use Management: The Economics of Public Forestlands.* Washington, DC: Resources for the Future.

Brown, K. 2001. "Industry Hugs Biotech Trees." *Technology Review.* Available at: www.technologyreview.com/magazine/mar01/innovation6.asp. Accessed on April 30, 2003.

Bruce, A., and J. Palfreyman, eds. 1998. *Forest Products Biotechnology.* London: Taylor & Francis.

Brussard, P. F., D. D. Murphy, and R. F. Noss. 1992. "Strategy and Tactics for Conserving Biological Diversity in the United States." *Conservation Biology* 6: 157–159.

Cabarle, B., R. Hrubes, C. Elliot, and T. Synnott. 1995. "Certification Accreditation: The Need for Credible Claims." *Journal of Forestry* 93 (4): 12–16.

Callicot, B. 1991. "Conservation of Biologic Resources: Responsibility to Nature and Future Generations." In D. J. Decker, M. E. Krasny, G. R. Groff, C. R. Smith, and D. W. Gross, eds., *Challenges in the Conservation of Biological Resources: A Practitioner's Guide.* Boulder, CO: Westview Press.

Chicago Climate Exchange. 2003. Available at: http://www.chicagoclimatex.com. The founding members of the Chicago Climate Exchange are American Electric Power (AEP); Baxter International Inc.; City of Chicago; Equity Office Properties Trust; Ford Motor Company; International Paper; Manitoba Hydro; MeadWestvaco Corporation; Motorola Inc.; STMicroelectronics; Stora Enso North America; Temple-Inland Inc.; and Waste Management Inc.

Clawson, M. 1967. *The Federal Lands since 1956: Recent Trends in Use and Management.* Baltimore, MD: Johns Hopkins University Press for Resources for the Future.

———. 1975. *Forests for Whom and for What?* Baltimore, MD: Johns Hopkins University Press for Resources for the Future.

———. 1983. *The Federal Lands Revisited.* Baltimore, MD: Johns Hopkins University Press for Resources for the Future.

Coulombe, M. 1995. "Sustaining the World's Forests: The Santiago Agreement." *Journal of Forestry* 93 (4): 18–21.

Daily, G. 1997. *Nature's Services: Societal Dependence on Natural Ecosystems.* Washington, DC: Island Press.

Daily, G., and K. Ellison. 2002. *The New Economy of Nature: The Quest to Make Conservation Profitable.* Washington, DC: Island Press.

Dana, S. T. 1918. "Forestry and Community Development." *USDA Bull.* 638. Washington, DC: U.S. Department of Agriculture.

Ehrlich, P. R. 1997. "No Middle Way on the Environment." *The Atlantic Monthly* 280 (6): 98–104.

FAO. 2000. *Global Outlook for the Future Wood Supply from Forest Plantations.* Rome: United Nations Food and Agriculture Organization.

———. 2001. *State of the World's Forests 2001.* Rome: United Nations Food and Agriculture Organization.

Faustmann, M. 1849. "The Determination of Value Which Forest Land and Immature Stands Possess for Forestry." Originally published in German in *Allegemeine Forest und Jagd Zeitung* 25. English translation published in 1995 in Umea, Sweden, in *Journal of Forest Economics* 1 (1).

Friedman, S. 2002. "Overview of Current Laws and Regulatory Structure." In *Biotech Branches Out: A Look at the Opportunities and Impacts of Forest Biotechnology.* Washington, DC: Pew Initiative on Food and Biotechnology.

Gordon, J. 1994. *The New Face of Forestry: Exploring a Discontinuity and the Need for a New Vision.* Pinchot Distinguished Lecture. Milford, PA: Grey Towers Press.

Gray, G. J., M. J. Enzer, and J. Kusel. 1998. *Understanding Community-Based Forest Ecosystem Management: An Editorial Synthesis of an American*

Forests Workshop, Bend, OR, June 1998. Washington, DC: American Forests.

Grumbine, M. 1994. "What Is Ecosystem Management?" *Conservation Biology* 8 (1): 27–38.

Heske, F. 1938. *German Forestry.* New Haven, CT: Yale University Press.

Hirt, P. 1994. *A Conspiracy of Optimism: Management of the National Forests since World War Two.* Lincoln: University of Nebraska Press.

Hoberg, George. 1999. "The Coming Revolution in Regulating Our Forests." *Policy Options.* Institute for Research in Public Policy. 53–56.

Howard, J. L. 1999. *U.S. Timber Production, Trade, Consumption and Price Statistics 1965–1997.* Gen. Tech. Rep. FPL-GRT-116. Madison, WI: U.S. Department of Agriculture, U.S. Forest Service, Forest Products Laboratory.

Howard, S., and J. Stead. 2001. *The Forest Industry in the 21st Century.* London: World Wide Fund for Nature.

Hunter, M. L., and A. Calhoun. 1996. "A Triad Approach to Land-Use Allocation." In R. A. Szaro and D.W. Johnston, eds., *Biodiversity in Managed Landscapes.* London: Oxford University Press.

International Union for the Conservation of Nature (IUCN). 1988. *Plant Conservation Programme.* Gland, Switzerland: IUCN.

Krutilla, J., and J. Haigh. 1978. "An Integrated Approach to National Forest Management." *Environmental Law* 8 (2).

Langbein, W., ed. 1996. *Seventh American Forest Congress: Final Report,* New Haven, CT: Yale School of Forestry and Environmental Studies.

Lee, J. 2003. "Seven States to Sue EPA over Standards on Air Pollution." *New York Times*, February 21.

Leopold, A. 1949. *A Sand County Almanac, and Sketches Here and There.* New York: Oxford University Press.

Mater, C. M., W. C. Price, and V. A. Sample. 2002. *Certification Assessments on Public and University Lands: A Field-Based Comparative Evaluation of the Forest Stewardship Council and Sustainable Forestry Initiative Programs.* Washington, DC: Pinchot Institute for Conservation.

MeadWestvaco. 2002. "Chicago Climate Exchange Founding Members Make Unprecedented Voluntary Binding Commitment to Reduce Greenhouse Gas Emissions by Four Percent by 2006." Available at: www.meadwestvaco.com/news/news011603.html. Accessed on April 30, 2003.

Meridian Institute. 2001. *Comparative Analysis of the Forest Stewardship Council and Sustainable Forestry Initiative Certification Programs.* Washington, DC: Meridian Institute.

Owusu, R. "GM Technology in the Forest Sector: A Scoping Study for WWF." Available at: http://www.wwf-uk.org/news/gm.pdf. Accessed on April 14, 2003.

Pagiola, S., J. Bishop, and N. Landell-Mills. 2002. *Selling Forest Environmental Services: Market-Based Mechanisms for Conservation and Development.* London: Earthscan Publications Ltd.

Pianin, E. 2003. "Reductions Sought in Greenhouse Gases." *Washington Post,* January 9, p. A4.

Pinchot, G. 1947. *Breaking New Ground.* New York: Harcourt, Brace and Company.

Prasad, B., ed. 1999. *Biotechnology and Biodiversity in Agriculture/Forestry.* Plymouth, England: Science Publishers Inc.

Pyne, S. J. 1982. *Fire in America: A Cultural History of Wildland and Rural Fire.* Seattle: University of Washington Press.

———. 2001. *Year of the Fires: The Story of the Great Fires of 1910.* New York: Viking.

Rathe, T. 1992. "The Green Area of the Green Market: Is It Really Environmentally Friendly? Solutions to Confusion Caused by Environmental Advertising." *Journal of Corporate Law* 18 (2): 419–458.

Raup, D. M., and J. J. Sepkoski. 1984. "Periodicity of Extinctions in the Geologic Past." *Proceedings of the National Academy of Sciences* 81: 801–805.

Raven, P. H. 1987. "We're Killing Our World: Preservation of Biological Diversity." *Vital Speeches of the Day,* May 15, 472–478.

Raychaudhuri, S., and K. Maramorosch, eds. 1999. *Biotechnology and Plant Protection in Forestry Science.* Plymouth, England: Science Publishers Inc.

Ricketts, T. H., E. Dinerstein, D. M. Olson, and C. J. Loucks. 1999. *Terrestrial Ecoregions of North America: A Conservation Assessment.* Covelo, CA: Island Press.

Sample, V. A. 1990. *The Impact of the Federal Budget Process on National Forest Planning.* Westport, CT: Greenwood Press.

———. 1998. *Principles of Sustainable Forest Management: Examples from Recent US and International Efforts.* Discussion Paper 98–01. Washington, DC: Pinchot Institute for Conservation.

Schallau, C. 1974. "Forest Regulation II: Can Regulation Contribute to Economic Stability?" *Journal of Forestry* 72 (6): 214–216.

Schmidheiny, S. 1992. *Changing Course: A Global Business Perspective on Development and the Environment.* Cambridge, MA: MIT Press.

Sedjo, R., and D. Botkin. 1997. "Using Forest Plantations to Spare Natural Forests." *Environment* 39 (10): 15–20.

Sedjo, R., and B. Sohngren. 2000. *Forestry Sequestration of CO_2 and Markets for Timber.* Discussion Paper 00–35. Washington, DC: Resources for the Future.

Shands, W. E., V. A. Sample, and D. Le Master. 1990. *National Forest Planning: Searching for a Common Vision.* Washington, DC: U.S. Department of Agriculture, U.S. Forest Service.

Smith, W., J. Vissage, R. Sheffield, and D. Darr. 2001. *Forest Resources of the United States, 1997.* General Technical Report NC-219. St. Paul, MN: U.S. Department of Agriculture, U.S. Forest Service, North Central Forest Experiment Station.

Society of American Foresters. 1995. "Forest Certification. Summary of Report of SAF Study Group on Forest Certification." *Journal of Forestry* 93 (4): 6–10.

Spies, T. A., and J. F. Franklin. 1996. "The Diversity and Maintenance of Old-Growth Forests." In R. A. Szaro and D. W. Johnston, eds., *Biodiversity in Managed Landscapes.* London: Oxford University Press.

State of Washington v. Avery Dexter. 1949. 32 Wn. 2d 551, 202 P. 2d 906.

Stedfast, S. 1999. "Regulatory Takings: A Historical Overview and Legal Analysis for Natural Resource Management." *Environmental Law* (Winter).

Szaro, R. A., and B. Shapiro. 1990. *Conserving Our Heritage: America's Biodiversity.* Arlington, VA: The Nature Conservancy.

Thomas, J. W. 1994. *Ecosystem Management: A National Framework.* Washington, DC: U.S. Department of Agriculture, U.S. Forest Service.

UNFCCC. 1992. "United Nations Framework Convention on Climate Change." Convention text available at http://unfccc.int/resource/docs/convkp/conveng.pdf. Accessed on April 30, 2003.

———. 1997. *Kyoto Protocol to the United Nations Framework Convention on Climate Change.* Protocol text available at http://unfccc.int/resource/docs/convkp/kpeng.pdf. Accessed on April 30, 2003.

U.S. Environmental Protection Agency. 1993. *Status Report on the Use of Environmental Labels Worldwide.* EPA 742-R-9–93–001. Washington, DC: U.S. Environmental Protection Agency.

USDA, U.S. Forest Service. 1999. "National Forest System Roadless Areas: Notice of Intent to Prepare an Environmental Impact Statement." *Federal Register* 64 (201): 56306–56307 (October 19, 1999).

USDA, U.S. Forest Service. 2001. "36 CFR Part 294. Special Areas: Roadless Area Conservation; Final Role." *Federal Register* 66 (9): 3244–3273 (January 12, 2001).

Waggener, T. R. 1977. "Community Stability as a Forest Management Objective." *Journal of Forestry* 75 (11): 710–714.

Westoby, J. 1989. *Introduction to World Forestry: People and Their Trees.* Oxford, England: Blackwell.

Williams, M. 1989. *Americans and Their Forests: A Historical Geography.* Cambridge, England: Cambridge University Press.

Wilson, E. O. 1992. *The Diversity of Life.* New York: W. W. Norton and Co.

World Commission on Environment and Development. 1987. *Our Common Future.* New York: Oxford University Press.

World Wildlife Fund. 2001. "Top Ten Companies Can Help Save the World's Forests." Press release, March 14. Washington, DC: World Wildlife Fund.

3

Chronology

1626   In the first documented forest legislation in what later would become the United States, the Massachusetts Bay Colony addresses local concerns over dwindling timber supplies by prohibiting the sale or transport of any timber whatsoever out of the colony without the approval of the governor and council.

1691   England reserves to the Crown "all pine trees fit for masts" throughout the American colonies; this Broad Arrow policy (named for the distinctive mark of the British Navy stamped into the best-suited white pines) establishes the right of the sovereign to exercise absolute control over the use of public lands. Widely unpopular with colonists, the policy is among the grievances contributing decades later to the American Revolution.

1781   Creation of the "original public domain" of Federal lands begins with the cession of "western reserve lands" by the 13 original states; further additions to the public domain are made through the Louisiana Purchase (1803), Florida Purchase (1819), Oregon Compromise (1846), Mexican Cession (1848), Texas Purchase (1850), Gadsden Purchase (1853), and Alaska Purchase (1867).

1799   Congress authorizes the creation of naval timber reserves to protect stands of live oak and red cedar needed for ship construction, reflecting the right of government to exercise control over public lands in the national interest.

1837   Senator John Calhoun of South Carolina introduces a bill providing for the cession to the states of all the public lands within their borders; although the proposal fails, it helps lead to the Preemption Act of 1841, in which 500,000 acres of Federal land is ceded to each of the existing states and to such new states as might be admitted to the Union.

1850   First of several railroad land grants are made to provide rights-of-way, assist railroads in the financing of railway construction, and promote settlement of the western United States; transcontinental grants are made in 1862 to the Union Pacific and Central Pacific Railroads, giving title to alternating 640-acre blocks of Federal land for 20 miles on each side of the railroad over its entire length. Eventually totaling more than 91 million acres, these grants remain the largest conveyances of public land to private corporations in the history of the nation.

1853   The warship USS *Michigan* is called on to control violent mobs in Manistee, Michigan, and assist a local agent of the new U.S. Department of the Interior in arresting several individuals caught stealing large quantities of timber from the public lands. Timber theft from public lands elsewhere in the country is widespread and largely uncontrolled.

1862   The Homestead Act allows an individual citizen to claim up to 160 acres of land from the public domain and to secure a free patent on that land upon proof of having resided on and cultivated the land for five years.

1864   Publication of George Perkins Marsh's *Man and Nature* (republished in 1872 as *Earth as Modified by Human Ac-*

tion), describing the gradual destruction of the Mediterranean forests through overexploitation and raising the prospect of similar destruction of America's abundant forests; this seminal work eventually helps prompt a reexamination of forest policy in the United States.

1872 The Mining Law of 1872 opens all public-domain lands to mineral exploration and purchase by private citizens when valuable mineral deposits are located and claimed. It is largely on the basis of this law that private citizens and corporations continue to own inholdings in Federal forests, which can be developed and to which transportation access over adjacent public land must be allowed.

1873 Timber Culture Act authorizes grants of 160 acres of Federal land per person to anyone who agrees to plant 40 acres of the land to trees not more than 12 feet apart and to keep them in a growing and healthy condition by cultivation for a period of ten years. This is intended to address the lack of wood for settlements in the western plains and prairies.

1877 Interior Secretary Carl Schurz introduces a timber-agent program to vigorously enforce laws against the theft of timber from the public lands; Congress sharply reduces Interior appropriations available for prosecution of such theft.

1878 Congress passes the Free Timber Act, which together with subsequent amendments permits the free cutting of timber anywhere on the public domain not only for mining, agricultural, and domestic purposes but also for manufacturing. In the same year, the Timber and Stone Act allows for the purchase of Federal timberlands in Oregon, Washington, California, and Nevada for $2.50 an acre, setting off a wave of fraudulent land purchases resulting in more than 15 million acres of the most productive timberlands in the Pacific Northwest being transferred from the public domain to private timber companies for a fraction of their value.

1879   Public dissatisfaction with fraud and abuse associated with public land acquisition leads to creation of the Public Land Commission, which recommends that all public lands be withdrawn from sale or other disposal; the commission recommends the sale of timber from these lands at fair market value and that settlers be allowed free use of timber for building, agricultural, mining, and other purposes, but not for sale, commerce, or export; Congress takes no action on the recommendations.

1885   New York becomes the first state to create state Forest Reserves, followed by Pennsylvania in 1897.

1891   Forest Reserve Act authorizes the President to reserve forest lands from the public domain for retention in federal ownership. It also repeals the Timber Culture Act and halts the sale of Federal lands except isolated small tracts. By the end of 1893, more than 17 million acres of Forest Reserve have been set aside by executive order, but no further action is taken because of the lack of any provision for the protection and administration of the reserves.

1897   Organic Administration Act directs that the Forest Reserves will be managed by the Department of the Interior to protect the headwaters of navigable waterways and to secure a continuous supply of wood for the American people; act specifies forest management standards that will later be interpreted as a prohibition of clearcutting on Federal forest lands.

1905   Second American Forest Congress held at the White House, organized by Gifford Pinchot and hosted by President Theodore Roosevelt; lays political groundwork for the transfer of the Forest Reserves from the Department of the Interior to the Department of Agriculture's Division of Forestry in the Transfer Act of 1905; Forest Reserves renamed *National Forests*; U.S. Forest Service established as an agency of U.S. Department of Agriculture, with Gifford Pinchot serving as first Chief Forester.

1906   Annual lumber production in the United States reaches a record 46 billion board feet; with the Lake States and the East largely cut over, most timber cutting has moved to the virgin forests of the Pacific Northwest and the South.

1910   President William H. Taft fires Gifford Pinchot as Chief Forester after Pinchot challenges Interior Secretary Richard A. Ballinger's approval of allegedly fraudulent claims on coal lands in the Chugach National Forest in Alaska. After a lengthy public dispute in which Congress is finally obliged to intervene, Pinchot is dismissed for providing information directly to Congress without approval by the Secretary of Agriculture as required by a standing executive order. National press characterizes Pinchot as a martyr for conservation in the fight against big business.

1911   Weeks Law provides long-sought authority for federal acquisition of forest lands, ostensibly to facilitate interstate commerce through the regulation of stream flow and protection of the headwaters of navigable rivers. This becomes the basis for the acquisition, with required state government approval, of most of what are now National Forests in the eastern United States.

1913   Underwood Tariff Act eliminates virtually all import tariffs and duties on wood products, settling the first of many trade disputes with Canada over timber trade. This dispute began in 1872 with the imposition of a lumber import duty by the United States. Canada responds with its own duty on the export of logs, severely restricting timber supply to American sawmills in the Lake States. Compromises in 1890 (McKinley Tariff Act) and 1894 (Wilson Act) are overturned by the U.S. forest products industry in the Dingley Tariff Act of 1897, which starts a new round in the trade dispute and prompts Canadians to ban all exports of logs from Crown lands to the United States. The Underwood Tariff Act compromise holds until 1930, when the cycle of U.S.-Canadian lumber trade disputes begins again.

1915   Forest Service is authorized to issue special-use permits for the construction of summer cabins on Federal lands, recognizing recreation as a legitimate use of the National Forests.

Branch of Research is established in the Forest Service as an organizational coequal of the branch administering the National Forests; regional experiment stations are established around the country, largely as a result of pressure from private timberland owners on Congress.

1916   National Park Service is established under Stephen T. Mather "to conserve the scenery and the natural and historic objects and the wildlife therein and to provide for the enjoyment of the same in such manner and by such means as will leave them unimpaired for the enjoyment of future generations."

1920   The Capper Report (prepared by the Forest Service at the request of Republican Senator Arthur Capper from Kansas) describes destructive cutting methods and fires following logging as having left 81 million acres of U.S. forest land effectively devoid of forest growth, increasing timber prices and raising concerns over future timber shortages. The transfer of much of the more productive public forest land to private ownership, followed by exploitive and unstable forest ownership (referred to as "cut-out-and-get-out") stimulates policies encouraging reforestation and a general shift from logging virgin forests to practicing sustained-yield management in second-growth forests.

National Forestry Program Committee is formed to secure the adoption and enactment of a comprehensive Federal forestry policy "to coordinate the various plans and measures proposed to provide an adequate and permanent timber supply for the people of the United States, and to promote national legislation to this end." Subsequent legislation focuses on the issue of Federal regulation of forestry practices on private lands. Prevailing view is that Forest Service should provide national leadership and financial assistance, but that regulation should be left to state governments under the police powers granted to states under the Constitution.

1924   Clarke-McNary Act establishes Federal-state cooperative forest fire control and cooperation in reforestation and management of state and private forests. Responsibility for forest fire control and forest pest management on all but the Federal forest lands eventually resides almost entirely with state government.

National Conference on Outdoor Recreation is convened under President Calvin Coolidge to determine how to address the growing public demand for forest-based recreation. A permanent committee established by Coolidge to implement the conference's resolutions focuses primarily on transfers from National Forests and other public lands to create new National Parks.

1930   Knutson-Vandenberg Act authorizes the Secretary of Agriculture to require a special deposit by purchasers of National Forest timber to help cover the expenses of reforestation and other mitigation measures on cutover areas. The objective is to speed up reforestation and improve silvicultural practices in the National Forests.

Shipstead-Nolan Act withdraws from entry all public lands in two counties in northern Minnesota, taking the first steps toward the later designation of roadless areas and wilderness preserves. This act requires the Forest Service to preserve the natural beauty of the lakes in the Superior National Forest primarily for recreational use and prohibits the cutting of all trees along the lakes' shorelines. This large protected area forms the basis for the later designation of the Boundary Waters Canoe Area and the establishment in cooperation with Canada of the Quetico-Superior International Peace Park.

1944   Sustained Yield Forest Management Act authorizes
            the secretaries of the Departments of Agriculture and
            the Interior to enter into cooperative agreements with
            private landowners to create "cooperative sustained-
            yield units" consisting of both Federal and private for-
            est lands. The purpose of these units is to promote the
            stability of forest industries, employment, communi-
            ties, and the taxable forest wealth through a continu-
            ous and stable supply of timber from the combination
            of Federal and private lands in the vicinity.

1948   Federal Water Pollution Control Act and subsequent
            major amendments require states to develop pro-
            grams for reducing and controlling water pollution,
            including nonpoint-source pollution from forestry ac-
            tivities such as sedimentation from road-building ac-
            tivities, fertilizer runoff, and spraying of pesticides;
            despite certain silvicultural exemptions, limitations
            on the draining and filling of inland wetlands be-
            comes an important consideration in forestry activi-
            ties.

1955   Clean Air Act and subsequent major amendments re-
            quire states to develop programs for reducing and
            controlling air pollution; visibility protection require-
            ments for Federal "Class I" areas become an impor-
            tant consideration in forest fire control and the use of
            prescribed burning for brush control and ecological
            restoration.

1960   Multiple-Use Sustained-Yield Act establishes that the
            National Forests are to be managed for a combination
            of uses, including outdoor recreation, range, timber,
            watershed, and wildlife and fish. *Multiple use* is de-
            fined as "the management of all the various renew-
            able surface resources of the National Forests so that
            they are utilized in the combination that will best meet
            the needs of the American people . . . with considera-
            tion being given to the relative values of the various
            resources, and not necessarily the combination of uses
            that will give the greatest dollar return or the greatest
            unit output." *Sustained yield* is defined as the
            "achievement and maintenance in perpetuity of a

high level annual or regular periodic output of the various renewable resources of the National Forests without impairment of the productivity of the land." This law becomes the foremost guiding principle for management, planning, and future policy development for the National Forests.

1964   The Wilderness Act provides for areas of Federal lands to be designated by law (rather than simply by the administrative authority of Federal land management agencies) and administered "for the use and enjoyment of the American people in such manner as will leave them unimpaired for future use and enjoyment as wilderness." The National Wilderness Preservation System (NWPS) is created from lands already under the management of at least four Federal agencies (Forest Service, National Park Service, Bureau of Land Management, and Fish and Wildlife Service), and management of individual segments of the NWPS continues to reflect the policies of those individual agencies.

1965   Land and Water Conservation Fund Act provides for federal acquisition of land, particularly for outdoor recreation, using a fund built up primarily from a portion of federal receipts from oil and gas leases on the outer continental shelf.

1970   National Environmental Policy Act requires an assessment of the potential environmental effects of any major federal action, a consideration of a full range of alternatives to the proposed action, and an opportunity for public involvement prior to final decision-making. Subsequently, environmental analysis becomes a major consideration in federal forest management and is the basis for many proposed activities being halted by the courts or significantly modified.

1973   Endangered Species Act provides for the protection of habitat for federally listed threatened or endangered species and prohibits development or forest management activities that would amount to a "taking" of these species through destruction of their habitat.

1974   Eastern Wilderness Act designates areas of several eastern National Forests as Wilderness and creates additional Wilderness Study Areas for future consideration. This is the start of a process for the review of relatively undeveloped areas of Federal forests for potential Wilderness designation that pits development interests against Wilderness protection interests and places forestry agencies in the middle of a continuing, politically charged debate.

Forest and Rangeland Renewable Resources Planning Act requires the periodic assessment of trends and conditions on all forests and rangelands in the United States and the development of a strategic plan for the management of renewable resources on National Forest lands.

1976   National Forest Management Act requires the periodic development of a management plan for each individual National Forest subject to several new limitations on controversial forest practices such as clearcutting and requiring public involvement.

1980   The Alaska National Interest Lands Conservation Act establishes several new National Monuments to protect portions of the 17-million-acre Tongass National Forest in southeast Alaska but legislatively establishes the timber harvest level; this sets up conflicts with the Endangered Species Act and other laws that keep the planning and management for this—the largest of the National Forests—tied up in the Federal courts for decades.

1988   Major forest fires in Yellowstone National Park and elsewhere in the western United States prompt a major reconsideration of long-standing fire control policies, which are seen as allowing an unnatural and dangerous accumulation of fuels and circumstances in which large-scale wildfires erupt with unacceptable loss of life and property; new policies are developed to allow natural fires to burn uncontrolled in remote forest areas and to expand the use of prescribed fire to reduce brush and other fuels in managed forests adjacent to developed areas.

1989   Northern spotted owl listed as a threatened species (under terms of the Endangered Species Act and National Forest Management Act); this results in the withdrawal of large areas of Federal forest in the western United States from timber harvesting; subsequent sharp reductions in timber harvest volume and forest industry employment, especially in the Pacific Northwest and California, result in one of the twentieth century's biggest controversies in forest management. Legal challenges substantially increase the role of the Federal courts in U.S. forest policy.

Several European countries ban the import of tropical timbers as way to slow the rate of tropical deforestation; as a result, the Forest Stewardship Council is formed by representatives of forest industry and conservation nongovernmental organizations to set standards by which a forest management enterprise can become independently certified and its products labeled as having come from well-managed forests; this is the start of several voluntary, market-based initiatives aimed at giving consumers the information needed to orient their purchases toward products from sustainably managed forests; it also represents an important step beyond sole reliance on government policy and regulation as a means of influencing the management of forests.

1990   Sharply accelerated international consolidation of the forest products industry results in the sale of large areas of private forest land for development; public concern over the loss of these forest areas, particularly in the northeastern United States, prompts new Federal laws to facilitate state government acquisition of private forest lands of significant conservation value and encourages the transfer of development rights on private forest lands to local government or qualified nongovernmental organizations.

1992   Global attention is focused on the rapid rate of defor-
            estation in the Amazon and tropical forests elsewhere
            in the world, with widespread implications of defor-
            estation for rural poverty, the loss of biodiversity, and
            acceleration of global climate change; a United Na-
            tions Conference on Environment and Development
            is held in Rio de Janeiro, resulting in an international
            convention on biological diversity, a framework con-
            vention on climate change, and, a nonbinding agenda
            for actions by both developed and developing coun-
            tries to improve the sustainable management of
            forests.

1993   United States takes part in a nonbinding international
            agreement to periodically assess domestic forest con-
            ditions and trends with reference to a set of seven cri-
            teria and 67 indicators of sustainable forest manage-
            ment (Montreal Process Criteria and Indicators, or
            C&I); in subsequent international meetings, United
            States participates in additional nonbinding agree-
            ments to take actions to gradually improve the man-
            agement of domestic forests with reference to the C&I.

1994   The two remaining long-term timber sale contracts for
            harvesting timber on the Tongass National Forest in
            southeast Alaska are canceled early as no longer being
            in the public interest; the long-term contracts, guaran-
            teeing access to Federal timber for 50 years, are the
            last of several such contracts begun in the 1950s to
            stimulate economic development in the region.

1995   Through a rider on the Forest Service appropriations
            bill for 1996, all timber salvage sales in the National
            Forests are made temporarily exempt from legal chal-
            lenge or judicial review under any law. Environmen-
            tal opposition is galvanized, and efforts to build con-
            sensus and cooperation among U.S. forest interests,
            such as the Seventh American Forest Congress, suffer
            a major setback.

1999   The United States and 46 other nations agree to the
            Kyoto Protocol, which commits countries to gradually
            reduce their emission of carbon dioxide and other

greenhouse gases through reduced fossil fuel consumption and the increased sequestration of atmospheric carbon through the use of "carbon sinks" such as forests; the United States officially withdraws from the agreement in 2001, but private entrepreneurs move ahead with the development of financial exchanges that allow carbon-emitting industries to purchase "carbon credits" to support afforestation and other carbon-storing activities to offset their emissions.

2000   The United Nations Forum on Forests is formed to oversee the implementation of more than 200 nonbinding "proposals for action" by individual countries that have agreed to conduct comprehensive nationwide assessments of the condition of their forests and steps they are taking to improve their management.

2002   Large-scale wildfires during the summers of 2000 and 2001 over many regions of the United States, especially in western states, raise the issue to the level of a White House initiative to reduce fuel buildups and fire hazards in Federal forests. Broad political support for addressing the forest health issue is tempered by public concerns that this will be used to sharply increase the level of commercial timber harvesting on Federal forests.

2003   Congress grants the Forest Service and the Bureau of Land Management authority to enter into long-term stewardship contracts for land management services to restore degraded forest ecosystems and address forest conditions conducive to destructive wildfires.

4

Personalities

The list of people who have made significant contributions to the development of U.S. forest policy is long, and those whose names are included in this chapter are but a few from among this group. The emphasis here is on individuals associated with major changes in U.S. forest policy and those mentioned elsewhere in the text of this book. These sketches provide a brief background and help the reader who seeks more detailed biographical information in other listed sources.

Bolle, Arnold (1912–1993)

As dean of the University of Montana School of Forestry in 1970, Arnold Bolle was asked by Democratic Senator Lee Metcalf of Montana to appoint a team of faculty members to conduct an independent investigation of alleged abusive forest management practices in the Bitterroot National Forest. The resulting "Bolle Report" became a focus for the debate over clearcutting in the national forests, culminating six years later with the National Forest Management Act.

Church, Frank (1924–1984)

During the 24 years that Democrat Frank Church served as U.S. senator from Idaho (1957–1981), he made several landmark contributions to U.S. forest policy. In the wake of numerous lawsuits over Forest Service clearcutting in the national forests, Senator Church held hearings in the Senate Interior Subcommittee on

Public Lands that produced the nonbinding "Church Subcommittee Guidelines" to limit the use of this practice. The guidelines were adopted as policy by the Forest Service and Bureau of Land Management and later served as a basis for the National Forest Management Act. Senator Church was also a strong advocate for Wilderness protection in Idaho, particularly the 2.2-million-acre River of No Return Wilderness, to which his name was added after his death.

Dana, Samuel Trask (1883–1978)

As an educator and first dean of the University of Michigan School of Natural Resources, Samuel Trask Dana was both a leader in and chronicler of the development of U.S. forest policy. He chaired the U.S. Timber Conservation Board and served on the National Outdoor Recreation Review Commission. He was also a member of the Hoover Commission on reorganization of the executive branch of government, which among other things recommended removing the Forest Service and the National Forests from the Department of Agriculture and consolidating them with all other natural resource functions in the Department of the Interior. This has since been attempted several times, but never successfully.

Dwyer, William (1929–2002)

As a Federal judge in the Ninth Circuit Court of Appeals, William Dwyer presided over more than a decade of legal challenges to the management of Federal forests in the Pacific Northwest. His decisions over such issues as the adequacy of Forest Service protection of old-growth forest habitat for the northern spotted owl had a profound effect on the forests and economy of the Pacific Northwest.

Fernow, Bernhard (1851–1923)

Born in Germany, Bernhard Fernow emigrated to the United States in 1876 and became a leader in the movement to protect forests against fire and exploitation. Fernow was one of the first professional foresters to introduce European concepts of reforestation and sustained-yield forest management to the United States. In 1886 he was appointed Chief of the Division of Forestry in the U.S. Department of Agriculture. After a trip through the western United States in 1887, Fernow concluded that "the future of the country cannot be left wholly to the operation of enlightened self-interest" and advocated for government leadership in the conservation of natural resources. Working with Senator Hale of Maine, he drafted a bill providing for the establishment and management of Forest Reserves, which later became the basis for the Forest Reserve Act of 1891. He organized several of the schools of forestry in the United States and Canada, and served as the first dean of the forestry school at Cornell University.

Franklin, Jerry (1936–)

As a forest ecologist at the University of Washington, Jerry Franklin served as a leading authority on the ecological functioning of old-growth forests of the Pacific Northwest during a decade of debate over policies to protect these ecosystems and the threatened or endangered species that inhabit them, such as the northern spotted owl. Franklin served as one of the "Gang of Four"—a group of scientists whose analysis and recommendations provided the basis for old-growth habitat conservation plans and management policies in the Pacific Northwest.

Gordon, John (1939–)

A forest ecologist and dean of the Yale School of Forestry and Environmental Studies, John Gordon has made numerous contributions to forest policy through the application of ecosystem principles to resource management. Gordon served as one of the "Gang of Four," a group of scientists whose analysis and recommendations provided the basis for old-growth habitat conservation plans and management policies in the Pacific Northwest.

Graves, Henry (1871–1951)

Graves served as the first dean of the Yale School of Forestry until 1910, when he was asked to succeed the recently fired Gifford Pinchot as chief of the U.S. Forest Service. During his tenure as chief, Graves warded off efforts to devolve the national forests to state and local control, expanded National Forests to the East under the Weeks Law of 1911, and organized the research branch of the Forest Service. One of Henry Graves' most important accomplishments was development of the first comprehensive

Policy of Forestry for the Nation in 1919, which later served as the basis for the Clark-McNary Act of 1924. This act was broadly supported, and represented a broad consensus on forestry among government, industry, and public interests that has been difficult to achieve in more recent decades.

Hough, Franklin (1822–1885)

Franklin Hough was perhaps the single most influential individual in the establishment of the precursors to both the Forest Service and the National Forests. As a chairman of a committee of the American Association for the Advancement for Science, Hough was instrumental in the founding of the Division of Forestry in the U.S. Department of Agriculture and served as its first director. Hough compiled a series of comprehensive studies of the extent and conditions of forests in the United States in 1877, 1878, and 1882 and called for the establishment of forest reserves to halt the disposal of public lands and conserve forest resources under the protection of the Federal government. Hough's efforts helped lay the foundation for the Forest Reserve Act of 1891 and the establishment of reserves that later became the basis for the National Forest System.

Humphrey, Hubert (1911–1978)

Hubert Humphrey served as vice president of the United States (1965–1969), but it was as Democratic U.S. senator from Minnesota that he made his most important contributions to U.S. forest policy. He was the lead sponsor of the Forest and Rangeland Renewable Resources Planning Act of 1974, which required the Forest Service to do a periodic assessment of the conditions and trends in U.S. forests and a strategic plan for their conservation and management. After the 1974 lawsuit over clearcutting on the Monongahela National Forest (*Izaak Walton League v. Butz*), Humphrey strove to "get forestry out of the courts and back into the woods" as lead sponsor of the National Forest Management Act of 1976.

Ickes, Harold (1874–1952)

Serving as Secretary of Interior from 1933 to 1946, Harold Ickes brought about important changes in policies governing the National Parks and the public-domain lands. The Taylor Grazing Act helped to establish the Grazing Service, which was the precursor to today's Bureau of Land Management. The O&C Act established sustained-yield management in forests in southern Oregon, revested to the Federal government after the failure of the Oregon and California Railroad. Ickes is perhaps best known for his concerted but unsuccessful attempt over several years to have the Forest Service and the national forests transferred from the Department of Agriculture to a revamped Department of the Interior, to be called the Department of Conservation.

Johnson, K. Norman (1942–)

A professor of forest engineering and operations research, K. Norman Johnson was a leading proponent of the use of optimization models in forest planning (FORPLAN) to balance the competing resource demands on the national forests. Johnson was one of the "Gang of Four," a group of scientists whose analysis and recommendations provided the basis for old-growth habitat conservation plans and management policies in the Pacific Northwest.

Leopold, Aldo (1886–1948)

Aldo Leopold is best known for his articulation of the individual's personal responsibility for land stewardship that maintains the land's productivity and ecological integrity—what he termed a "land ethic." Leopold was a forester with the U.S. Forest Service and later a professor of game management at the University of Wisconsin. His 1949 book *A Sand County Almanac* is regarded as a major inspiration for later policies aimed at managing land as an integrated ecosystem, rather than as a collection of parts, and recognizing the interdependence of the many plant and animal species that define ecosystems.

Marsh, George Perkins (1801–1882)

As author of the 1864 book *Man and Nature*, George Perkins Marsh helped Americans recognize that continued unfettered exploitation of forests and other natural resources could lead to ecological disaster. As American ambassador to Italy and an amateur naturalist, Marsh was shocked to realize that much of the barren, unproductive land around the Mediterranean had once been forests and fertile cropland, but had been damaged by centuries of overuse. His scientific approach and his advocacy for more sustainable resource use helped launch the Conservation Movement in the United States.

Marshall, Robert (1901–1939)

Bob Marshall was one of the leading proponents of a policy of protecting certain undeveloped areas of the Federal lands as Wilderness. He served as the first director of forestry for the U.S. Bureau of Indian Affairs, and later as the head of the Forest Service's Division of Recreation and Lands, where he was instrumental in protecting more than 5 million acres of wilderness. One of the founders of the Wilderness Society, he was commemorated with the naming of the Bob Marshall Wilderness Area in Montana, one of the nation's largest Wilderness areas.

Mather, Stephen (1867–1930)

The founder of the National Park Service, Stephen Mather sought permanent protection of forest areas of outstanding scenic, historic, and recreational value on Federal lands. Now comprising more than 80 million acres nationwide, the national parks have become worldwide models for protected-area management and scientific study, and they educate millions of visitors annually about the value of protecting wild landscapes.

McGuire, John (1916–2002)

While serving as chief of the Forest Service during the turbulent 1970s, John McGuire faced increasing opposition to forestry practices being carried out on the National Forests. Congress held hearings on clearcutting on the National Forests, brought about by lawsuits on the Bitterroot National Forest in Montana and the Monongahela National Forest in West Virginia. As a result, Congress mandated planning at the National Forest, regional, and national levels through the Resources Planning Act of 1974 and the National Forest Management Act of 1976. McGuire ushered in important changes in forest management practices and methods of land management.

Muir, John (1838–1914)

Founder of the Sierra Club, John Muir is credited with introducing many Americans to the beauty and grandeur of America's forests and mountains and the need for national policies to ensure their protection from development. Along with President Theodore Roosevelt and Forest Service Chief Gifford Pinchot, Muir was a powerful advocate for conservation and the creation of National Parks, but he differed with Roosevelt and Pinchot over permitting the regulated use of Federal lands for activities such as grazing and water development.

Murie, Olaus (1889–1963)

A lifelong advocate of protecting certain wild forest areas as Wilderness, Olaus Murie led the decades-long campaign for the creation of a National Wilderness Preservation System, which was accomplished with the Wilderness Act of 1964.

Olmsted, Frederick Law (1822–1903)

As one of the premier landscape architects of his time, Frederick Law Olmsted established forests as an essential part of even the urban landscape and helped create an appreciation for natural beauty that later translated to broad support for natural resource conservation. Known primarily for his creation of New York City's Central Park, Olmsted was also a leading advocate for the protection of wild areas such as Yosemite Valley.

Pack, Charles Lathrop (1857–1937)

One of the most colorful and controversial figures in American forest policy, Charles Lathrop Pack accumulated a substantial fortune logging the virgin forests of the Pacific Northwest. As president of the American Forestry Association during the early twentieth century, Pack became an advocate for forest conservation, research, and education, ultimately endowing experimental forests at several U.S. universities and establishing a foundation that continues to support forestry education.

Peterson, Max (1927–)

Max Peterson led the U.S. Forest Service as Chief during a difficult time of turmoil and criticism of the agency. Major forest policy issues included the development of National Forest plans implementing the National Forest Management Act of 1976, the identification of remaining National Forest roadless areas for possible Wilderness designation, the effect of the recession and the housing slump of the early 1980s on timber-dependent communities, increasing public concern over the use of herbicides and pesticides on the National Forests, growing concern about old-growth logging, below-cost timber sales, and threatened and endangered species. During this period, the Forest Service's funding was sharply reduced, and public trust of the Forest Service to effectively manage the National Forests fell to its lowest point in the agency's history.

Pinchot, Gifford (1865–1946)

Regarded by many as the "Father of Forestry in the United States," Gifford Pinchot served as the first Chief Forester and head of the U.S. Forest Service. As a leader of the Progressive Era Conservation Movement, Pinchot's fight for conservation focused on the introduction of scientific forestry to the United States and the establishment of national policies to promote the conservation and sustainable use of forests on both public and private lands. With the 1905 transfer of the Forest Reserves from the Department of the Interior to the Forest Service in the Department of Agriculture, Pinchot established the overarching policy that these National Forests were always to be managed "for the greatest good, for the greatest number, in the long run." Pinchot also founded several other nongovernmental institutions to promote forest science and sound forest management, including the Yale Forest School and the Society of American Foresters.

Powell, John Wesley (1834–1902)

John Wesley Powell was a leading advocate for reform of Federal land policies aimed at promoting settlement of the western United States. His geological exploration of the interior West, and particularly the Colorado River watershed, helped persuade policymakers that a different approach was needed in the development of these arid lands. As director of the U.S. Geological Survey, he initiated the first major biological and geological surveys of the West and helped establish the role of the Federal government in advancing science as a basis for natural resource policy.

Roosevelt, Theodore (1858–1919)

As twenty-sixth president of the United States, Theodore Roosevelt helped to firmly establish the Progressive Era Conservation Movement, basing it on the notion that the nation's natural resources were a public trust rather than an asset to be exploited for the benefit of a few private individuals or corporations. In 1905, President Roosevelt signed the Transfer Act, which established the National Forests and the U.S. Forest Service as we know it today. In cooperation with Gifford Pinchot, his Chief Forester, President Roosevelt increased the area of these Federal Forest Reserves from 56 million acres to 148 million acres. According to Roosevelt, "The nation behaves well if it treats the natural resources as assets which it must turn over to the next generation increased and not impaired in value."

Sargent, Charles Sprague (1841–1927)

A botanist, Charles Sargent was appointed by President Grover Cleveland to head the Forestry Commission, which the National Academy of Sciences had formed to advise on the protection of forest land in the United States. The commission's recommendations, urged forward by its vice chairman Gifford Pinchot, resulted in President Cleveland's establishment of the first 13 Forest Reserves.

Schenck, Carl (1868–1955)

Founder of the first forestry school in the United States, a private school on George Vanderbilt's Biltmore Estate in North Carolina, Carl Alwin Schenck compiled the first state-by-state analysis of forest area and ownership, distribution of forest types, and forest conservation laws in effect, and he later published them as *Lectures in Forest Policy* (1904).

Thomas, Jack Ward (1934–)

Just prior to being named Chief of the Forest Service in 1992, Jack Ward Thomas was the senior scientist on several major studies concerning the protection of the threatened northern spotted owl and old-growth habitat in the Pacific Northwest. Thomas was a member of the "Gang of Four," a group of scientists whose analysis and recommendations provided the basis for old-growth habitat conservation plans and management policies in the Pacific Northwest. In 1993, Thomas was named to head the Forest Ecosystem Management Assessment Team (FEMAT), which had as its objective to resolve the spotted owl crises in the Pacific Northwest and northern California based on the best scientific evidence.

Udall, Morris (1922–1988)

As a congressman and chair of the House Interior Committee, Mo Udall was one of the strongest modern-day advocates for conservation in the U.S. government. He strove to find balanced solutions to controversial issues over the management of Federal forests, water, and Wilderness. He played a major role in the Alaska National Interest Lands Conservation Act, the Tongass Timber Reform Act, the Strip Mining Reclamation Act, and the Southern Arizona Water Rights Settlement Act.

Weyerhaeuser, Frederick (1834–1912)

Beginning with 900,000 acres of Washington forests purchased from James J. Hill and the Northern Pacific Railway, Frederick Weyerhaeuser founded the Weyerhaeuser Timber Company. The company eventually owned more than 6 million acres of forest land in the United States, and for many years it was the largest and most politically influential private forestry firm in the country.

Weyerhaeuser, Frederick E. (1872–1945)

President of the Weyerhaeuser Company from 1928 until his death, F. E. Weyerhaeuser was an industry leader in adopting the policy of sustained-yield management of commercial timberlands—a policy that had previously been applied primarily to public forests. He was a proponent of prompt planting and reforestation following timber harvesting.

Zahniser, Howard (1906–1964)

President of the Wilderness Society from 1945 to 1964, Howard Zahniser is regarded as the father of the Wilderness Act, which was passed in 1964 after 66 drafts spanning almost 20 years.

Zon, Raphael (1887–1957)

As an early pioneer in the development of Forest Service research, Raphael Zon was the author of what became a landmark study on the importance of forests in watershed protection, *Forests and Water in the Light of Scientific Investigation* (1927).

5

Forest Facts and Data

The forests of the United States are diverse in their character, ownership, and management. U.S. forests range from tropical in Hawaii, southern Florida, and Puerto Rico to the boreal spruce-fir forests of Alaska and northern Maine. The forests are a mix of conifer (or evergreen) and deciduous (or broadleaf) species. The forests grow right down to the ocean's edge and up to the timberlines of the Colorado Rocky Mountains. The United States is the largest consumer of forest products in the world but harvests far less per year than what its forests are capable of growing per year.

Who owns the forest? The Federal government owns a large amount of forest land, but private citizens own more. Comparatively, the forest industry owns a fairly small amount of forest land. Tribal governments, states, counties, and municipalities own the rest.

This chapter presents facts, statistics, and other information intended to provide the reader with a basic understanding of U.S. forests—what they are, how they are managed, who owns and manages them, and their relative condition. The chapter also describes two international initiatives that affect U.S. forest management: (1) the Montreal Process Criteria and Indicators and (2) Forest Management Certification. For more in-depth information on any of these subjects, readers should consult the information resources enumerated in chapter 7.

Forest Ecoregions in the United States

An ecoregion is a geographic area that has a relatively distinct combination of biotic and abiotic factors, such as soils, vegetation, climate, geology, physical geography, elevation, and latitude. A forest ecoregion is a geographic area that has a relatively distinct combination of tree species as well as other biotic and abiotic factors. The most cited source for delineation of U.S. ecoregions is a map produced in 1995 by Robert G. Bailey and his colleagues (http://www.fs.fed.us/land/ecosysmgmt/ecoreg1_home.html).

The United States has four major domains: Dry, Humid Tropical (Hawaii, Puerto Rico, and the Florida Everglades), Humid Temperate, and Polar (Alaska) (Bailey 1995). The vast majority of the forest land in the United States is in the Humid Temperate domain. Within each domain are divisions, and within each division are provinces corresponding to geographic location, climate, and dominant vegetation types.

In general, conifer forests such as pine, true fir, Douglas fir, hemlock, larch, false cedar, redwood, sequoia, and spruce are found in Alaska, the western United States, the northern Lake States, the northern tier of New England, and the piedmont and coastal plains of the Southeast and in Texas.

Deciduous forests—those with oak, maple, beech, cherry, walnut, poplar, ash, birch, and aspen—are found primarily in the Lake States, the mid-South, the mid-Atlantic, the southern tier of New England, and Appalachia. In many regions, conifer and deciduous forests mix, making it difficult to draw clean boundaries between forest ecoregions.

Forest types have evolved and persisted in response to soil and climate conditions as well as because of latitudinal and longitudinal location. In many parts of the United States, large disturbances have changed the entire forest composition. For example, the chestnut blight has eliminated the American chestnut from the forest ecosystem in the eastern United States. Early land clearing for farming eliminated much of the native deciduous forest in New England and the Ohio River valley. Today New England is heavily forested with many of same forest tree species that were present before European settlement, but the forested areas are interspersed with roads and development. By contrast, the Ohio River valley is still heavily fragmented with farms, urban centers, and forest. Aggressive logging in Michigan and Minnesota during westward migration greatly reduced the red and eastern white pine forests. In their place are extensive stands of aspen, birch, Jack pine, and scattered—but growing—pockets of red and eastern white pine. Human settlement and natural forces such as disease, wind, and fires have shaped the forests' ecoregions over time.

An excellent printed publication of ecoregions of the United States is Ricketts et al. (1999), *Terrestrial Ecoregions of North America: A Conservation Assessment.*

Forest Area, Production, and Trade Information

Information for this section came from the following sources: Cubbage, O'Laughlin, and Bullock, *Forest Resource Policy* (1993); U.S. Forest Service, *RPA Assessment of Forest and Range Lands* (2000); and U.S. Forest Service, *U.S. Forest Facts and Historical Trends* (2001).

In the United States, 747 million acres (33 percent of the total land area) is forest land. Timberland accounts for 504 million acres, or 67 percent of the total forest land area. Of the remaining forest land area, 51.9 million acres is designated as *reserved forests* (for example, designated Wilderness areas), and 190.2 million acres is designated *other forests*—forests not capable of producing commercial timber on an annual basis for whatever reason.

Timberland is the production part of the forested land. The southern United States has the most available timberland with 200 million acres, followed by the northern United States with 160.5 million acres and the western United States with 143.3 million acres. Conversely, the western United States contains 76 percent of reserved forest (39.5 million acres); and the northern and southern United States have 7.4 million acres and 4.9 million acres, respectively, of reserved forest.

U.S. timber production and timber trade can be summarized by the following statistics:

- The United States produces almost 30 percent of the world's pulp, 33 percent of the world's paper and cardboard, and 20 percent of the world's softwood lumber.
- Since 1980, the United States has harvested about

450 million cubic meters of timber annually. Of this total, softwood harvest (used primarily for construction lumber and paper and cardboard pulp) comprises 64 percent or 285 million cubic meters, and hardwood harvest (used primarily for furniture) comprises 36 percent or 165 million cubic meters.

- Forest growth exceeds forest harvest by a ratio of 3:2 in public forests. On privately owned forest industry land, removal significantly exceeds growth, with a ratio of 5:1.
- Timber harvested from public forests makes up about 17 percent of average annual total harvest (80 million cubic meters out of 450 million cubic meters).
- On average, the United States consumes 33 percent more lumber than it produces domestically. Canadian imports make up the majority of this deficit.
- The United States exports about 13 percent of the world's pulp, 7 percent of the world's paper and cardboard, and less than 5 percent of the world's softwood lumber. Canada is the largest exporter of these three products.
- The southern United States is by far the region with the highest harvest levels, at about 290 million cubic meters annually or 64 percent of average annual total harvest, followed by the western and northern United States, respectively.

Ownership of U.S. Forests

There are numerous owners of U.S. forests, most of them falling under the "private nonindustrial" ownership class (Table 5.1). The term *nonindustrial* means that no wood processing mill is associated with the forest properties. Private nonindustrial owners—typically individuals, families, trusts, and some corporations—own 49 percent of the nation's forests. Despite the political and economic importance of the forest industry, this ownership class only controls 9 percent of the total forest land in the United States. The Federal government owns and manages about 33 percent of U.S. forest lands under different mandates, ranging from multiple-use to strictly Wilderness. Of the Federal public lands, the U.S. Forest Service owns and manages 60 percent, followed by the Bureau of Land Management and other Federal agencies such as National Park Service, Department of Defense, and Bureau of Indian

Table 5.1   Area of Forest Land and Timberland in the United States by Ownership Category, 1997

	Acres in millions	
Ownership category	Forest land*	Timberland†
Federal	246.7	109.2
National Forest System	146.8	96.4
Bureau of Land Management	48.3	6.1
Other	51.6	6.6
State	60.5	28.9
County and Municipal	9.2	7.9
Private	430.5	357.7
Forest Industry	67.7	66.9
Nonindustrial Private Forests	362.8	290.8
TOTAL	746.9	503.7

Source: W. B. Smith, J. S. Vissage, D. R. Darr, and R. M. Sheffield. 1997. *Forest Resources of the United States, 1997.* Gen. Tech. Rep. NC-219. St. Paul, MN: USDA Forest Service, North Central Research Station.

* Forest land is land at least 10 percent stocked by forest trees of any size, including land that formerly had tree cover and can be naturally or artificially regenerated.
† Timberland is land that is producing or is capable of producing crops of industrial wood and has not been withdrawn from timber production by statute or administrative regulation.

Affairs. States, counties, and municipalities own about as much as the forest industry owns—about 9 percent of all U.S. forests.

However, not all U.S. forest lands are suitable for producing timber. The U.S. Forest Service has separated timberlands from forest lands. The distribution of timberland ownership is identical to the distribution of forest land ownership.

Federal Public Forest System

There are four general categories of Federal public forest lands: the National Forest System (NFS), Bureau of Land Management (BLM) lands, the National Park System (NPS), and the National Wilderness Preservation System (NWPS). The first and last categories are managed by the U.S. Forest Service; the middle two categories are managed by agencies within the U.S. Department of the Interior. National Forests and BLM lands not designated as Wilderness are, for the most part, managed as multiple-use. That is, these forests are intended to produce a variety of commercial and noncommerical goods, benefits, services, and values, such as game and nongame wildlife, recreation, timber, grazing, mineral exploration, oil and gas, water development, biodiversity, and subsistence uses (food, shelter, and medicine).

National Parks and the Wilderness Preservation System lands are managed for preservation, thereby eliminating virtually all commercial uses, especially timber harvesting. These lands are also intended to allow natural ecological processes including wildfire to occur unimpeded. The Yellowstone National Park fires of 1988 were intentionally allowed to burn. By contrast, many fires on the National Forests and BLM forest lands are actively suppressed. All Federal public forest lands receive their funding from annual congressional appropriations, and recreation-use fees also contribute to operating budgets.

The National Forest System (www.fs.fed.us)

The National Forest System (NFS) includes National Forests and National Grasslands located in 44 states, Puerto Rico, and the Virgin Islands. Established by the Creative Act of 1891, today's National Forest System has 155 National Forests and 20 National Grasslands totaling 191 million acres—8.5 percent of the total land in the United States. Managing these 191 million acres is the U.S. Forest Service, which is part of the U.S. Department of Agriculture. The Forest Service comprises the Washington, D.C., office, 9 regional offices, 124 supervisors' offices, and more than 600 administrative units called ranger districts. Each level of the agency has different legal mandates, operational procedures, and resource management responsibilities.

The primary policies governing NFS land and resource management are the Organic Administration Act of 1897, the Multiple-Use Sustained-Yield Act of 1960, the Wilderness Act of 1964, the National Environmental Policy Act of 1969, the Endangered Species Act of 1973, and the National Forest Management Act of 1976.

The nine regional offices (numbered 1–6, 8, 9, and 10) are:

- *Region 1:* Twelve National Forests in northern Idaho, northeastern Washington, and Montana; four National Grasslands in North Dakota and northwestern South Dakota. Contact information:

  USDA Forest Service
  Northern Region
  200 E. Broadway
  P.O. Box 7669
  Missoula, MT 59807
  Phone: (406) 329-3511

- *Region 2:* Eleven National Forests in Colorado, Nebraska, South Dakota, and Wyoming; four National Grasslands in Colorado, Nebraska, and Wyoming. Contact information:

  USDA Forest Service
  Rocky Mountain Regional Office
  740 Simms Street
  P.O. Box 25127
  Lakewood, CO 80225-0127
  Phone: (303) 275-5350

- *Region 3:* Eleven National Forests in Arizona and New Mexico; two National Grasslands in Oklahoma and the Texas Panhandle. Contact information:

  USDA Forest Service
  Southwestern Region
  333 Broadway SE
  Albuquerque, NM 87102
  Phone: (505) 842-3292

- *Region 4:* Thirteen National Forests in Utah, western Wyoming, southern Idaho, Nevada, and small portions of California and Colorado. Contact information:

  USDA Forest Service
  Intermountain Region
  324 25th Street
  Ogden, UT 84401
  Phone: (801) 625-5306

- *Region 5:* Seventeen National Forests in California and the Lake Tahoe Basin Management Area. Contact information:

  USDA Forest Service
  Pacific Southwest Region
  1323 Club Drive
  Vallejo, CA 94592
  Phone: (707) 562-8737

- *Region 6:* Nineteen National Forests in Oregon and Washington; one National Grassland in Oregon. Contact information:

USDA Forest Service
Pacific Northwest Region
P.O. Box 3623
333 S.W. First Avenue
Portland, OR 97208-3623
Phone: (503) 808-2971

- *Region 8:* Thirty-five National Forests in Alabama, Arkansas, Florida, Georgia, Kentucky, Louisiana, Mississippi, North Carolina, Oklahoma, Puerto Rico, South Carolina, Tennessee, Texas, and Virginia; two National Grasslands in Texas; Land Between the Lakes National Recreation Area in Kentucky; Savannah River Site in South Carolina. Contact information:

USDA Forest Service
Southern Region
1720 Peachtree Road NW
Atlanta, GA 30309
Phone: (404) 347-4177

- *Region 9:* Fourteen National Forests in Illinois, Indiana, Michigan, Minnesota, Missouri, New Hampshire, New York, Ohio, Pennsylvania, Vermont, West Virginia, and Wisconsin; Midewin National Tallgrass Prairie in Illinois. Contact information:

USDA Forest Service
Eastern Region
319 W. Wisconsin Avenue, Suite 500
Milwaukee, WI 53203
Phone: (414) 297-3600

- *Region 10:* Two National Forests in Alaska. Contact information:

USDA Forest Service
Federal Office Building
709 W. 9th Street
P.O. Box 21628
Juneau, AK 99802-1628
Phone: (907) 586-8806

Bureau of Land Management (www.blm.gov)

The Bureau of Land Management (BLM) is responsible for managing 262 million acres of land, or about one-eighth of the land in the United States. Additionally, about 300 million acres of subsurface mineral resources are managed by the BLM. The bureau is also responsible for wildfire management and suppression on 388 million acres.

The land managed by the BLM is located in 12 western United States including Alaska. Much of this land is dominated by extensive grasslands, forests, high mountains, arctic tundra, and deserts. Management activities of BLM land include energy and minerals; timber; forage; wild horse and burro populations; fish and wildlife habitat; Wilderness areas; archaeological, paleontological, and historical sites; and other natural heritage values.

The 12 state offices are:

Bureau of Land Management
Alaska State Office
222 W. 7th Avenue #13
Anchorage, AK 99513
Phone: (907) 271-5960

Bureau of Land Management
Arizona State Office
222 N. Central Avenue
Phoenix, AZ 85004
Phone: (602) 417-9200

Bureau of Land Management
California State Office
2800 Cottage Way
Suite W1834
Sacramento, CA 95825-1886
Phone: (916) 978-4400

Bureau of Land Management
Colorado State Office
2850 Youngfield Street
Lakewood, CO 80215
Phone: (303) 239-3600

Bureau of Land Management
Eastern States Office
7450 Boston Boulevard
Springfield, VA 22153-3121
Phone: (703) 440-1600

Bureau of Land Management
Idaho State Office
1387 S. Vinnell Way
Boise, ID 83709
Phone: (208) 373-4000

Bureau of Land Management
Montana/Dakotas State Office
5001 Southgate Drive
P.O. Box 36800
Billings, MT 59107
Phone: (406) 896-5000

Bureau of Land Management
Nevada State Office
1340 Financial Boulevard
P.O. Box 12000
Reno, NV 89520
Phone: (775) 861-6400

Bureau of Land Management
New Mexico State Office
1474 Rodeo Road
Santa Fe, NM 87505
Phone: (505) 438-7400

Bureau of Land Management
Oregon State Office
333 S.W. 1st Avenue
Portland, OR 97204
Phone: (503) 808-6002

Bureau of Land Management
Utah State Office
P.O. Box 45155
324 S. State Street
Salt Lake City, UT 84145-0155
Phone: (801) 539-4001

Bureau of Land Management
Wyoming State Office
5353 Yellowstone Road
P.O. Box 1828
Cheyenne, WY 82003-1828
Phone: (307) 775-6256

National Park System (www.nps.gov)

The National Park System (NPS) encompasses approximately 83.6 million acres nationwide. The largest area is Wrangell–St. Elias National Park and Preserve, Alaska. At 13.2 million acres, it comprises 16.3 percent of the entire system. The smallest unit in the system is Thaddeus Kosciuszko National Memorial, Pennsylvania, at 0.02 of an acre. The NPS includes National Parks, monuments, preserves, historic sites, historical parks, memorials, battlefields, cemeteries, recreation areas, seashores, lakeshores, rivers, parkways, and trails.

The National Wilderness Preservation System (www.nwps.gov)

There are more than 620 areas officially designated *Wilderness* in the National Wilderness Preservation System (NWPS). These areas total 104,571,344 acres, or 4.52 percent of the entire land area of the United States. Four agencies manage Wilderness lands: Bureau of Land Management (5 percent of NWPS); USDA Forest Service (33.2 percent of NWPS); U.S. Fish and Wildlife Service (19.8 percent of NWPS); and National Park Service (42.1 percent of NWPS). The western states (excluding Utah) and Alaska hold the largest number of acres of Wilderness. Each state has an average of 1.5 million acres. Only Connecticut, Delaware, Iowa, Kansas, Maryland, and Rhode Island have no Wilderness areas.

State and County Forest Lands

States own about 60.5 million acres of forest land, and counties and municipalities combined own 9.2 million acres of forest land in the United States (USDA, U.S. Forest Service 2000). State forest lands were either granted as trust lands on entry into the United States or were specifically designated to protect them from development—the 2.5-million-acre Adirondack State Park, for example. State-owned forest lands are managed for a variety of objectives. For example, Adirondack is managed for preservation goals and objectives, with no timber harvesting. State forests in Colorado, Washington, and Oregon, by contrast, are managed to produce revenue primarily for schools. For a more in-depth history and analysis of the management of state forest and other trust lands, see Souder and Fairfax, *State Trust Lands: History, Management, and Sustainable Use* (2002).

County and municipal forest lands are most often found in the Northeast and the Lake States. Many of these lands were acquired during the early part of the 1900s when homesteaders had to forfeit their lands because they could not pay taxes on them—so-called "tax-delinquent lands." Very little is known about county and municipal forest management.

Private Forest Lands

Private forest land accounts for 430 million acres or 58 percent of the total U.S. forest land. Private forest land is classified as forest industry land or nonindustrial private land (USDA, U.S. Forest Service 2000). Nonindustrial forest land ownership comprises 9.9 million units—primarily individuals, but also organizations, partnerships, and so on (Birch 1994). Forest industry lands are obviously managed for wood production; nonindustrial forest lands are managed for reasons ranging from investment or second homes to wildlife recruitment and family heirloom.

Forest Industry

The forest industry owns about 67 million acres of U.S. forest land, all of which is timberland. The southern United States accounts for the largest holdings of forest industry forest land at 54.7 percent, followed by the North at 22.1 percent, the Pacific Coast at 18.8 percent, and the Rocky Mountains at 4.3 percent. Ten percent of the hardwood growing stock and 14 percent of the softwood growing stock is held by the forest industry.

Timber Investment Management Organizations

The timber investment management organization (TIMO) is a relatively novel approach to timberland management in the United States—an emerging form of forest land ownership that fits between traditional industrial and nonindustrial categories. TIMOs (see the web site http://www.endgame.org/timo.html for additional information) are primarily made up of financial institutions that invest in private timberland. In general, TIMOs own and manage forest lands as a low-risk investment on behalf of large institutional investors such as teachers' pension funds. TIMOs emphasize steady rates of return over long time periods. Forests managed by TIMOs are not as intensively managed as industry lands, but they are more intensively managed than the average nonindustrial forest land. In 1998, investments totaled about $6.5 billion in more than 6 million acres of forest land.

The major TIMOs include the John Hancock Timber Resource Group (which owns the largest number of holdings), the Campbell Group, Forest Investment Associates, Forestland Group, Prudential Timber Investments, Strategic Timber Trust, Timberland Investment Services, UBS Resource Investments, U.S. Timberlands, Wachovia Timberland Investment Management, Wagner Forest Management Ltd., and Woodlands Resource Management Group.

Nonindustrial Private Forests

Nonindustrial private forests (NIPFs) account for nearly half of U.S. forest lands, or 363 million acres. East of the Mississippi, more than two-thirds of the forest land is NIPF. In the western United States the majority of forests are federally owned. NIPFs supply about 49 percent of the timber harvested in the United States. NIPFs are also ecologically important because of where they are located on the landscape. Private forest lands were settled first and were the most productive in terms of sustaining the early homesteaders. Private lands tend to be closer to water, have more productive soils, and are generally more ecologically rich than public lands, which tend to be at higher elevations and have thinner soils and less water. Federal and state public forest lands were the marginal leftover lands that no one could settle.

Nonindustrial private forests are diverse in their biophysical characteristics, ownership, uses, and management. An indispensable resource on nonindustrial and industrial private forest landowners in the United States is by Thomas W. Birch, entitled *Private Forest-Land Owners of the United States* (1994, USDA, U.S. Forest Service, Northeastern Forest Experiment Station). This report analyzes data about private forest landowners from 1978 and 1994. The general trend is for private forest lands to be either divided into smaller tracts owned by an increasing number of individuals or consolidated into large tracts owned by a decreasing number of corporations. Table 5.2 details the number of private forest landowners and acres in each type of land ownership.

The trend toward smaller tracts owned by many individual nonindustrial private forest landowners is called "parcelization." Table 5.3 gives the data connected to parcelization, which has two general effects. The first is that within any region, the objectives, uses, and management activities of nonindustrial forest landowners is likely to be diverse and widely variable. Achieving ecological goals such as wildlife habitat conservation, water quality, and forest biodiversity is an enormous social and political challenge, as these goals typically require coordinated management across very large areas and across large numbers of owners. Coordinating these landowners and encouraging their cooperation is no easy task.

The second effect of parcelization is the potential for fragmentation of the forest. Nonindustrial forest landowners may decide to convert some or all of their parcels into nonforest cover—residential development or agriculture. A large number of small openings across a forested landscape not only fragments forest cover but also disrupts wildlife habitat and can damage water-

Table 5.2 Estimated Number of Private Forest Landowners and Acres Owned by Type of Ownership, 1978 and 1994

Ownership type	Number of owners (thousands)		Percentage of owners		Acres owned (millions)		Percentage of acreage	
	1978	1994	1978	1994	1978	1994	1978	1994
Individual	6,793	9,319	87.6	94.1	183.5	232.3	55.1	59.0
Partnership	484	289	6.2	2.9	35.8	29.7	10.7	7.5
Corporation	237	157	3.0	1.6	101.1	107.1	30.4	27.2
Other	246	136	3.2	1.4	12.7	24.3	3.8	6.2
TOTAL	7,757	9,902	100.0	100.0	333.1	393.4	100.0	100.0

Source: T. W. Birch. 1994. *Private Forest-Land Owners of the United States, 1994.* Radnor, PA: USDA, U.S. Forest Service, Northeastern Forest Experiment Station.

Table 5.3   Estimated Number of Private Forest Landowners and Acres Owned by Ownership Size Class, 1978 and 1994

Ownership size class (acres)	Number of private owners (millions)		Percentage of private owners	
	1978	1994	1978	1994
1–9	5,528	5,795	71.3	58.6
10–49	1,164	2,762	15.0	27.9
50–99	464	717	5.9	7.2
100–499	538	559	7.0	5.6
500–999	40	41	0.5	0.4
1000+	23	27	0.3	0.3
TOTAL	7,757	9,902	100.0	100.0

Source: T. W. Birch. 1994. *Private Forest-Land Owners of the United States, 1994.* Radnor, PA: USDA, U.S. Forest Service, Northeastern Forest Experiment Station.

shed functioning. Property tax incentives or forest land conservation easements have some effect in minimizing parcelization, but the economics of private forest land ownership can, in many parts of the country, force nonindustrial forest landowners either to cut down their trees to generate immediate financial gain or to convert their forest land into nonforest cover and land uses. Either way, the full spectrum of environmental services provided by forests can be permanently negatively affected.

As far as the primary reason for owning forest land is concerned, private owners vary greatly. Of course, industrial forest landowners are primarily interested in generating revenue for themselves or shareholders, although many industrial forest landowners also have a real estate division. The reasons for owning forest land among nonindustrial private owners range from "part of residence" to "land investment." Table 5.4 identifies the reasons for owning forest land.

Tribal Forests

Tribal forest land comprises 16 million acres on 214 reservations in 23 states. Approximately one-half of the forests are timberland and the remainder are woodland (forest with less than 5 percent crown cover by commercial timber species).

Forest management is essential to the economies of the surrounding communities. The Federal government "has a trust responsibility for managing Indian forests" that is accomplished through the Bureau of Indian Affairs (BIA) and tribal governments.

Table 5.4   Estimated Number of Private Forest Landowners and Acres Owned by Primary Reason for Owning Forest Land, 1994

Reason for owning	Number of owners		Acres of forest land	
	Thousands	*Percent (rank)*	*Millions*	*Percent (rank)*
Land investment	920.0	9.3 (5)	39.3	10.0 (3)
Recreation	874.5	8.8 (6)	37.9	9.5 (5)
Timber production	272.2	2.7 (9)	113.2	28.9 (1)
Farm and domestic use	816.4	8.3 (7)	35.8	9.1 (6)
Enjoyment of owning	1,392.4	14.1 (3)	28.7	7.3 (8)
Part of farm	1,189.8	12.0 (4)	38.6	9.8 (4)
Part of residence	2,641.5	26.7 (1)	32.6	8.2 (7)
Other	1,440.9	14.5 (2)	61.0	15.3 (2)
No answer	354.1	3.6 (8)	6.3	1.5 (9)
TOTAL	9,901.7	100.0	393.4	100.0

Source: T. W. Birch. 1994. *Private Forest-Land Owners of the United States, 1994.* Radnor, PA: USDA, U.S. Forest Service, Northeastern Forest Experiment Station.

Detailed information about tribal forest lands can be found in *An Assessment of Indian Forests and Forest Management in the United States*, a publication of the Intertribal Timber Council.

Conditions and Trends in U.S. Forests

The condition of forests in the United States is a topic of lively debate. Statistics show that the *quantity* of forest land in the United States has been relatively unchanged for 80 years, with the greatest threat to the loss of forests being development. However, there are wide-ranging opinions about the *quality* of U.S. forests and prognoses for the future. Listed here are brief descriptions of U.S. forest conditions and trends from three sources: the Forest Service's *2000 RPA Assessment of Forest and Range Lands*, the World Resources Institute's *Low-Access Forests and Their Level of Protection in North America* (2002), and Douglas W. MacCleery's *American Forests: A History of Resiliency and Recovery* (1996).

- Total forest land area in the United States has been relatively stable since the 1920s (USDA, U.S. Forest Service 2000). Recent changes in forest land coverage include a 25.1-million-acre loss of forest land to other uses between 1982 and 1997. The losses can be attributed to land use changes as land is removed from agriculture or private forests are cleared for development. On forest land some forest types have changed since 1800. In the northeastern and mid-Atlantic states, the American chestnut was once in great abundance, but the chestnut blight has made the species all but extinct. In the southern Appalachians, mid-South, and piedmont zones of the Southeast, mixed hardwood (deciduous) forests have been converted to agriculture or intensively managed pine plantations. Large areas of red and eastern white pine forests in northern Michigan, Wisconsin, and Minnesota have been fragmented and replaced with aspen and other deciduous forest types. In the western United States, fire suppression has altered large areas of Ponderosa pine forests by allowing diverse tree species to grow in underneath Ponderosa pine canopies. What were once pure Ponderosa pine forests are now mixed-conifer forests with a high risk of uncharacteristically intense fires.

- Only 6 percent of forest cover in the lower 48 states remains within large tracts of low-access, low-elevation forests (WRI, 2002). These large tracts are at least 320 square miles and are mostly undivided by roads and other access routes. These large tracts are essential for maintaining viable populations of threatened or endangered wildlife and biodiversity in general. Continued logging and road building in these areas threaten biodiversity.

- About 39 million acres of forest land in the National Forest System—and many more millions of acres on private lands—are at high risk from catastrophic wildfires. Many of these forests are overcrowded from years of poor logging practices and fire suppression, resulting in high mortality rates from bark beetle and other insect and disease outbreaks. High mortality rates—standing and fallen dead trees—result in excessive fuel buildups that can rapidly feed wildfires.

- About 40 percent of the large tracts of low-access forest land is considered strictly or moderately protected from development activities such as logging, road building, or other human activities. Much of this forest is under federal ownership and management, 30 percent of it within the National Forest System. These remaining large tracts of forests are the focus of controversy regarding logging and road construction. The latest policy initiative dealing with these forests was the Roadless Area Rule, proposed during the Clinton Administration.

- The amount of timberland in the United States has remained stable since the 1950s at two-thirds of the forest land. Hardwood volumes increased between 1952 and 1997. Softwood volumes increased between 1952 and 1977, declined for the next 10 years, but increased again until 1997. The development of fast-growing hybridized or genetically modified plantation tree species will likely contribute to total wood production in the United States. On average in the United States, growth exceeds removal for hardwoods and softwoods with a combined growth:removal ratio of 3:2 in 1996. Trends in harvesting of nontimber forest products are difficult to establish because of the dependency on habitat, information availability, and so on. The acreage of plantations planted has increased. Private forest lands are expected to decline by 2 percent over the next 50 years. Hardwood and softwood growing stocks are expected to increase in the future, and timber harvest on public land should remain stable.
- In the lower 48 states, more than 60 percent of the remaining large tracts of low-access forest are found within six states. In order of relevant forest area, they are Idaho, Montana, Washington, California, Wyoming, and Minnesota.
- Tree mortality rates range between 0.5 and 0.8 percent of the growing timber stock. In some areas the mortality rates are expected to increase. About 58 million acres are at risk of greater mortality rates because of insect and disease infestations. Drought and altered forest conditions in the western United States attributable to 100 years of fire suppression are most at risk for insect and disease epidemics. Examples include spruce budworm, mountain pine beetle, bark beetle, and dwarf mistletoe.
- Fire frequency is on the rise across the United States after 20 stable years between the 1950s and 1970s. Only the South is considered to be within historic fire frequencies. Drought conditions in the West and Northeast are likely to exacerbate the frequency of large

fires. About 73 million acres of forest land in the western United States is regarded as at risk for large, intense wildfires in the next decade. Aggressive forest thinning and prescribed fire programs can offset the frequency and intensity of fires in small areas, but it is unlikely that any program can receive the funding and legal authority necessary to address the fire issue on a sufficiently large scale.

• Exotic species are increasing in the United States; the highest concentrations are in the North and along the Pacific Coast. There are more than 4,500 exotic species in the nation, prompting many plant ecologists to regard the invasion of exotic plant species in the United States as one of the greatest threats to biological diversity. Exotic pests are also a major threat. The Asian long-horned beetle, for example, was introduced to the United States as larvae inside untreated wood shipping pallets from China. This pest was first found infesting trees in New York and, more recently, Chicago. The gypsy moth has been and continues to be a major threat to hardwoods in the eastern United States since its introduction from Europe into Massachusetts in 1869.

The Montreal Process

The Montreal Process is the product resulting from the Working Group on Criteria and Indicators for the Conservation and Sustainable Management of Temperate and Boreal Forests. The working group was formed in 1990 during the UN Convention on Sustainable Development in Montreal, Canada. Membership in the Working Group is voluntary and currently includes countries from both hemispheres: Argentina, Australia, Canada, Chile, China, Japan, Republic of Korea, Mexico, New Zealand, Russian Federation, United States, and Uruguay. These countries represent about 90 percent of the world's temperate and boreal forests in the Northern and Southern Hemispheres.

The Montreal Process Working Group agreed on a framework of criteria and indicators that provide member countries with a common definition of what characterizes sustainable management of temperate and boreal forests. The framework identifies seven criteria that are further defined by 67 associated indicators that are aspects of the criteria that can be identified or described.

Criterion 1 through 6

The following six criteria and associated indicators (quoted from http://www.mpci.org/criteria_e.html) characterize the conservation and sustainable management of temperate and boreal forests. They relate specifically to forest conditions, attributes or functions, and to the values or benefits associated with the environmental and socioeconomic goods and services that forests provide. The intent or meaning of each criterion is made clear by its respective indicators. No priority or order is implied in the alphanumeric listing of the criteria and indicators. Indicators for which most data are available are so identified. All others may require the gathering of new or additional data and/or a new program of systematic sampling or basic research.

Criterion 1: Conservation of Biological Diversity

Biological diversity includes the elements of the diversity of ecosystems, the diversity between species, and genetic diversity in species.

Indicators of Biological Diversity:
- *Ecosystem diversity*
  a. Extent of area by forest type relative to total forest area [most data available]
  b. Extent of area by forest type and by age class or successional stage
  c. Extent of area by forest type in protected area categories as defined by IUCN [International Union for Conservation of Nature] or other classification systems [most data available] [IUCN categories include strict protection, ecosystem conservation and tourism, conservation of natural features, conservation

through active management, landscape/
seascape conservation and recreation, and
sustainable use of natural ecosystems.]
d. Extent of areas by forest type in protected
areas defined by age class or successional stage
e. Fragmentation of forest types

- *Species diversity*
  a. The number of forest dependent species
  b. The status (threatened, rare, vulnerable,
  endangered, or extinct) of forest dependent
  species at risk of not maintaining viable
  breeding populations, as determined by
  legislation or scientific assessment [most data
  available]

- *Genetic diversity*
  a. Number of forest dependent species that
  occupy a small portion of their former range
  b. Population levels of representative species from
  diverse habitats monitored across their range

Criterion 2: Maintenance of Productive Capacity of Forest Ecosystems

[The productive capacity of a forest ecosystem entails
the components and processes necessary for a forest
ecosystem to renew itself in perpetuity. Ecosystem
components and process include nutrients, nutrient
cycling, water retention, leaf respiration, decomposi-
tion, regeneration, seed production, plant growth, and
plant mortality.]

Indicators of Productive Capacity:

a. Area of forest land and net area of forest land
available for timber production [most data
available]
b. Total growing stock of both merchantable and
non-merchantable tree species on forest land

available for timber production [most data available]

c. The area and growing stock of plantations of native and exotic species [most data available]

d. Annual removal of wood products compared to the volume determined to be sustainable [most data available]

e. Annual removal of non-timber forest products (e.g., fur bearers, berries, mushrooms, game), compared to the level determined to be sustainable

Criterion 3: Maintenance of Forest Ecosystem Health and Vitality

[Forests maintain their health and vitality by recovering from and adapting to a variety of disturbance forces, such as insect and disease infestations, fire, wind, landslides, floods, and so on. The same components and processes that contribute to the productive capacity of forest ecosystems also contribute to forest ecosystem health and vitality.]

Indicators of Health and Vitality:

a. Area and percent of forest affected by processes or agents beyond the range of historic variation, e.g., by insects, disease, competition from exotic species, fire, storm, land clearance, permanent flooding, salinisation, and domestic animals

b. Area and percent of forest land subjected to levels of specific air pollutants (e.g., sulfates, nitrate, ozone) or ultraviolet B that may cause negative impacts on the forest ecosystem

c. Area and percent of forest land with diminished biological components indicative of changes in fundamental ecological processes (e.g., soil nutrient cycling, seed dispersion, pollination) and/or ecological continuity (monitoring of functionally important species such as fungi, arboreal epiphytes, nematodes, beetles, wasps, etc.)

Criterion 4: Conservation and Maintenance of Soil and Water Resources

This criterion encompasses the conservation of soil and water resources and the protective and productive functions of forests.

Indicators of Conservation:

a. Area and percent of forest land with significant soil erosion

b. Area and percent of forest land managed primarily for protective functions, e.g., watersheds, flood protection, avalanche protection, riparian zones [most data available]

c. Percent of stream kilometers in forested catchments in which stream flow and timing has significantly deviated from the historic range of variation

d. Area and percent of forest land with significantly diminished soil organic matter and/or changes in other soil chemical properties

e. Area and percent of forest land with significant compaction or change in soil physical properties resulting from human activities

f. Percent of water bodies in forest areas (e.g., stream kilometers, lake hectares) with significant variance of biological diversity from the historic range of variability

g. Percent of water bodies in forest areas (e.g., stream kilometers, lake hectares) with significant variation from the historic range of variability in pH, dissolved oxygen, levels of chemicals (electrical conductivity), sedimentation or temperature change

h. Area and percent of forest land experiencing an accumulation of persistent toxic substances

Criterion 5: Maintenance of Forest Contribution to Global Carbon Cycles

[Forests convert atmospheric carbon into oxygen through photosynthesis and respiration. Forests also store an enormous amount of carbon in the soil, root systems, tree trunks, branches, and leaves. Young forests grow very rapidly, converting atmospheric carbon to oxygen, but young forests do not store as much carbon as old-growth forests.]

Indicators of Contribution:

a. Total forest ecosystem biomass and carbon pool, and if appropriate, by forest type, age class, and successional stages
b. Contribution of forest ecosystems to the total global carbon budget, including absorption and release of carbon (standing biomass, coarse woody debris, peat and soil carbon)
c. Contribution of forest products to the global carbon budget

Criterion 6: Maintenance and Enhancement of Long-Term Multiple Socio-economic Benefits to Meet the Needs of Societies

[Forests provide society with a wide range of goods and services, from basic food and shelter to jobs and recreational experiences. There is a mutual relationship between forests and people: Forests contribute to the wealth and well-being of people; in turn, people will ensure that the forests are not exploited and will contribute to the sustainability of the forests.]

Indicators of Long-Term Benefits:

- *Production and consumption*
  a. Value and volume of wood and wood products production, including value added through downstream processing [most data available]

b. Value and quantities of production of nonwood forest products
c. Supply and consumption of wood and wood products, including consumption per capita [most data available]
d. Value of wood and non-wood products production as percentage of GDP
e. Degree of recycling of forest products
f. Supply and consumption/use of non-wood products

- *Recreation and tourism*
  a. Area and percent of forest land managed for general recreation and tourism in relation to the total area of forest land
  b. Number and type of facilities available for general recreation and tourism in relation to population and forest area
  c. Number of visitor days attributed to recreation and tourism in relation to population and forest area

- *Investment in the forest sector*
  a. Value of investment, including investment in forest growing, forest health and management, planted forests, wood processing, recreation and tourism [most data available]
  b. Level of expenditure on research and development, and education
  c. Extension and use of new and improved technologies
  d. Rates of return on investment

- *Cultural, social, and spiritual needs and values*
  a. Area and percent of forest land managed in relation to the total area of forest land to protect the range of cultural, social and spiritual needs and values
  b. Non-consumptive use forest values

- *Employment and community needs*
  a. Direct and indirect employment in the forest sector and forest sector employment as a proportion of total employment

> b. Average wage rates and injury rates in major employment categories within the forest sector [most data available]
> c. Viability and adaptability to changing economic conditions, of forest dependent communities, including indigenous communities
> d. Area and percent of forest land used for subsistence purposes

Criterion 7

Criterion 7 and associated indicators (quoted from http://www.mcpi.org/criteria_e.html) relate to the overall policy framework of a country that can facilitate the conservation and sustainable management of forests. Included are the broader societal conditions and processes often external to the forest itself but which may support efforts to conserve, maintain, or enhance one or more of the conditions, attributes, functions, and benefits captured in criteria 1 through 6. No priority or order is implied in the listing of the indicators.

Criterion 7: Legal, Institutional, and Economic Framework for Forest Conservation and Sustainable Management

Indicators of Framework:

- *Extent to which the* **legal framework** *(laws, regulations, guidelines) supports the conservation and sustainable management of forests, including the extent to which it:*
  - a. Clarifies property rights, provides for appropriate land tenure arrangements, recognizes customary and traditional rights of indigenous people, and provides means of resolving property disputes by due process;
  - b. Provides for periodic forest-related planning, assessment, and policy review that recognizes

the range of forest values, including coordination with relevant sectors;

c. Provides opportunities for public participation in public policy and decision-making related to forests and public access to information;

d. Encourages best practice codes for forest management;

e. Provides for the management of forests to conserve special environmental, cultural, social, and/or scientific values.

- *Extent to which the* **institutional framework** *supports the conservation and sustainable management of forests, including the capacity to:*

  a. Provide for public involvement activities and public education, awareness and extension programs, and make available forest-related information;

  b. Undertake and implement periodic forest-related planning, assessment, and policy review including cross-sectoral planning and coordination;

  c. Develop and maintain human resource skills across relevant disciplines;

  d. Develop and maintain efficient physical infrastructure to facilitate the supply of forest products and services and support forest management;

  e. Enforce laws, regulations and guidelines.

- *Extent to which the* **economic framework** *(economic policies and measures) supports the conservation and sustainable management of forests through:*

  a. Investment and taxation policies and a regulatory environment which recognize the long-term nature of investments and permit the flow of capital in and out of the forest sector in response to market signals, non-market economic valuations, and public policy decisions in order to meet long-term demands for forest products and services;

       b. Non-discriminatory trade policies for forest products.

- *Capacity to* **measure and monitor** *changes in the conservation and sustainable management of forests, including:*
  - a. Availability and extent of up-to-date data, statistics, and other information important to measuring or describing indicators associated with criteria 1–7;
  - b. Scope, frequency and statistical reliability of forest inventories, assessments, monitoring and other relevant information;
  - c. Compatibility with other countries in measuring, monitoring and reporting on indicators.

- *Capacity to conduct and apply* **research and development** *aimed at improving forest management and delivery of forest goods and services, including:*
  - a. Development of scientific understanding of forest ecosystem characteristics and functions;
  - b. Development of methodologies to measure and integrate environmental and social costs and benefits into markets and public policies and to reflect forest-related resource depletion or replenishment in national accounting systems;
  - c. New technologies and the capacity to assess the socio-economic consequences associated with the introduction of new technologies;
  - d. Enhancement of ability to predict impacts of human intervention on forests;
  - e. Ability to predict impacts on forests of possible climate change.

The Forest Service and several states are in the process of drafting programmatic reports documenting how Montreal Process C&Is are to apply to forest lands and resources they oversee. The Montreal Process Criteria and Indicators web site is http://www.mpci.org/criteria_e.html.

Forest Certification Principles and Standards

This section outlines the various principles and standards under different forest certification systems. Forest certification is a small but growing concept in the United States. Forest certification emerged in the early 1990s with the development of the Montreal Process Criteria and Indicators. The idea behind forest certification is to provide financial incentives for timber producers to employ forest practices that sustain and enhance ecological and social forest values, such as biodiversity and community sustainability. Forests are certified by meeting certain standards and practices that vary by certification system, bioregion, and country. The certification is carried through to the end-use forest products such as lumber and paper, guaranteeing purchasers that the products are derived from forests that are managed sustainably. Several large private forest companies have received certification through protocols sanctioned by the Forest Stewardship Council (FSC). The state of Pennsylvania and the Navajo Indian Nation are also certified. However, widespread adoption of forest certification among U.S. forest companies has been slow.

The FSC is an international nongovernmental, nonprofit organization that has taken the lead on timber certification. In countries across the world, FSC educates third-party organizations on how to certify wood and establishes criteria for sustainable management. In the United States, only two organizations are approved by the FSC to certify wood—SmartWood and Scientific Certification Systems.

SmartWood

SmartWood certification of forest "sources" is based on field review using SmartWood's generic guidelines, or, when available, country or bioregional guidelines that have been drafted in consultation with local experts and organizations, often in collaboration with the FSC. SmartWood has been involved in the development of draft regional certification standards throughout the world. In all cases, these guidelines have been developed with help from stakeholders in each region, including the general public, local communities, professional foresters, ecologists, and social scientists. The guidelines are widely and publicly circulated for comment and are periodically revised based on comments received.

Forest management certification evaluates the practices of forest managers according to environmental, silvicultural, and social standards. If these standards are met, an operation is certified, and timber (or other nontimber forest products) harvested from that forest may be sold as SmartWood and FSC-certified. In general, candidate operations must meet the following broad principles:

1. Long-term security for the forest.
2. Maintenance of environmental functions, including watershed stability and biological conservation.
3. Sustained-yield forestry production.
4. Positive effect on local communities.
5. The existence of a system for long-term forest management planning, management, and monitoring (including a written forest management plan). In the case of plantations, SmartWood does not endorse the conversion of standing forests to tree plantations, but it will certify those that have been developed on previously deforested lands and/or that are a first step toward forest restoration.

Certification for Chain of Custody for SmartWood
Certification of companies marketing SmartWood products (for example, wholesalers, processors, retailers, and brokers) is granted after chain-of-custody audits confirm that certified wood is being used in certified product lines.

Categories of Source Certification for SmartWood
SmartWood sources are certified according to how closely they adhere to SmartWood principles and guidelines. Sources operating in very strict adherence to these principles, and having long-term data to support this, are classified as *sustainable*. Sources that can demonstrate a strong operational commitment to the principles and guidelines are classified as *well managed*.

Categories of Chain-of-Custody Certification
Chain-of-custody certification assures consumers that the certified items they buy were produced with certified wood or other certified nontimber forest products. SmartWood companies re-

ceive chain-of-custody certification according to whether all or some of their wood products come from certified SmartWood sources. An "exclusive" SmartWood company sells forestry products made only from wood from SmartWood or other FSC-endorsed sources. A "nonexclusive" SmartWood company sells products from both SmartWood and other FSC-endorsed certified sources and other noncertified origins.

Scientific Certification Systems

The forest certification process of Scientific Certification Systems (SCS) is built on the premise that an evaluation of forest products for purposes of establishing a basis for marketplace claims must necessarily focus on the management of the land from which those products originate. The evaluation process includes the following components:

1. Information supplied by the landowner that can be verified through observation and field sampling.
2. Collection of additional field data through sample-based field reviews conducted by an interdisciplinary evaluation team under contract to SCS.
3. A structured evaluation process based on sound decision science principles that focuses on timber resource sustainability, forest ecosystem maintenance, and financial and socioeconomic considerations.
4. Ongoing, periodic monitoring to assure continued adherence to long-term management plans and management practices in place at the time of the initial evaluation and to assure adequate tracking of the chain of custody of products from certified operations.
5. Chain-of-custody certification for all participants involved in the manufacture and distribution of certified products destined for markets in which the product bears the "certified" claim.

Operational steps include the following:

1. Determine the scope of the potential project
2. Conduct a preliminary evaluation
3. Execute the contract
4. Assemble the evaluation team

- Determine evaluation scope; collect and analyze data
- Consult with regional stakeholders
- Weight criteria and modify scoring guide (and documentation), as necessary

5. Assign numerical performance scores
6. Specify conditions and recommendations
7. Solicit and respond to client review comments
8. Solicit and respond to peer review comments

Chain-of-Custody Process

Valid chain-of-custody procedures are a key aspect of any credible product-labeling program. The purpose of these procedures is to ensure that the product bearing a label is, in fact, produced from certified sources or materials. The process for chain-of-custody certification normally involves the following operational steps:

1. The participant is asked to submit a summary of processing/sales operations outlining how operational procedures will incorporate chain-of-custody consideration.
2. SCS reviews the summary to ensure that all elements of the respective chain-of-custody standards are covered.
3. If the approach appears to be viable for the purposes of maintaining chain of custody, SCS recommends that the participant proceed with an on-site compliance audit.
4. The participant reviews and signs the chain-of-custody contract.
5. An audit is scheduled.
6. On-site compliance audits are conducted by an SCS inspector to accomplish the following:
   - To ensure that the client's documentation is complied with by staff
   - To review activities and documentation to determine that the documentation of activities is sufficient
   - To review the effectiveness of the system in meeting chain-of-custody requirements
7. An audit report is submitted to the client for review.
8. A certificate is issued if the client's operations successfully meet chain-of-custody standards.

The Sustainable Forestry Initiative (SFI) (www.aboutsfi.org)

To meet the needs of both society's demand for wood products and ecological health, the SFI program was developed by foresters, conservationists, scientists, landowners, and other stakeholders. The SFI program allows foresters, landowners, loggers, and wood and paper producers to satisfy the growing demand for environmentally sound forestry while still producing forest products economically. An independent group of stakeholders called the Sustainable Forestry Board (SFB) manages the SFI program standard and verification procedures and SFI program compliance. Currently, the SFB has 15 members, two-thirds of whom represent nonindustry interest groups such as environmental and conservation groups.

The SFI program has 11 objectives or principles that provide guidance to forest managers who practice or want to practice sustainable forestry.

- Broadening the implementation of sustainable forestry by employing an array of economically, environmentally, and socially sound practices in the conservation of forests—including appropriate protection, growth, harvest, and use of those forests— using the best scientific information available.
- Ensuring long-term forest productivity and conservation of forest resources through prompt reforestation, soil conservation, afforestation, and other measures.
- Protecting the water quality in streams, lakes, and other bodies of water.
- Managing the quality and distribution of wildlife habitats and contributing to the conservation of biological diversity by developing and implementing stand- and landscape-level measures that promote habitat diversity and the conservation of forest plants and animals including aquatic fauna.
- Managing the visual impact of harvesting and other forest operations.
- Managing program participants' lands of ecologic, geologic, cultural, or historic significance in a manner that recognizes their special qualities.
- Promoting the efficient use of forest resources.

- Broadening the practice of sustainable forestry by cooperating with forest landowners, wood producers, consulting foresters, and program participants' employees who have responsibility in wood procurement and landowner assistance programs.
- Publicly reporting program participants' progress in fulfilling their commitment to sustainable forestry.
- Providing opportunities for the public and the forestry community to participate in the commitment to sustainable forestry.
- Promoting continual improvement in the practice of sustainable forestry and monitoring, measuring, and reporting performance in achieving the commitment to sustainable forestry.

American Tree Farm System Standards (http://www.treefarmsystem.org)

Since 1941, the American Tree Farm System (ATFS), under the oversight of the American Forest Foundation (AFF), has been certifying trees managed using sustainable forestry. It is the oldest and largest voluntary third-party verification process in the United States.

Certification in ATFS is a voluntary process and follows a set of standards and guidelines. Landowners who want to become certified Tree Farmers must have their property inspected by an ATFS forest professional. If the property meets AFF's standards and guidelines for forest sustainability, the landowner receives a certificate and the recognizable diamond-shaped Tree Farm sign. The property is reinspected every five years to maintain Tree Farm certification status. The inspections are available free of charge to the landowner.

Table 5.5 contains a list of the ATFS Standards, Guidelines, and Performance Measures for certification.

Through December 1997, close to 3.5 million acres of U.S. forest land has been certified by SmartWood, Scientific Certification Systems, or both (Hayward and Vetinsky 1999). SmartWood has certified 1,051,000 acres and SCS has certified 2,616,000 acres. The two organizations jointly certified more than 218,000 acres of the Menominee Tribal Enterprises. According to information on the Forest Stewardship Council web site, more than 30.4 million acres of forest land have been certified worldwide.

Table 5.5 American Tree Farm System (ATFS) Standards, Guidelines, and Performance Measures

Standards and guidelines for forest landowners	Performance measures for Tree Farm certification
Ensuring Sustainable Forests Standard: Members of the ATFS promote the growing of renewable forest resources on their forest land while protecting environmental benefits. They are encouraged to strive to increase public understanding of all benefits of productive forestry.	**Ensuring Sustainable Forests Performance Measures:** To achieve and maintain certification, all members must have a written and active forest management plan. This plan must take into consideration maintenance and/or enhancement of wood and fiber production, wildlife habitat, water quality, and recreational opportunities.
Reforestation Standard: Members must provide for prompt restocking of desired species of trees on harvested areas and idle areas where tree growing is the land use objective. Restocking may be accomplished by natural seeding, sprouting, direct seeding, or reforestation with tree seedlings.	**Reforestation Performance Measures:** To achieve and maintain certification, members must achieve satisfactory restocking levels within five years following harvest, or less than five years as specified by state or local ordinance. Acreage not reforested because of change of use shall be deducted from overall Tree Farm acreage.
Water Quality Standard: Forestry practices must include the application of the state's environmental protection organization–approved forestry best management practices (BMPs) or forest practices act as well as any other practices required by local, state, or federal regulations.	**Water Quality Performance Measures:** To achieve and maintain certification, members must be in compliance with state forestry best management practices (BMPs) or their forest practices act to ensure that water quality standards are met.
Wildlife Habitat Standard: Members' forest management plans must address the effects of forest practices on fish and wildlife.	**Wildlife Habitat Performance Measures:** To achieve and maintain certification, members shall follow forest practices that, to the extent practicable, protect and enhance fish and wildlife habitat while considering floral and faunal diversity.

(Table continues)

Table 5.5 American Tree Farm System (ATFS) Standards, Guidelines, and Performance Measures (continued)

Standards and guidelines for forest landowners	Performance measures for Tree Farm certification
Forest Aesthetics Standard: Members shall follow forest practices that consider the aesthetic effects of forest activities.	**Forest Aesthetics Performance Measures:** To achieve and maintain certification, members shall, to the extent practicable, follow forest management practices that demonstrate concern for visual impacts.
Protect Special Sites Standard: Implemented forest management practices shall, to the extent practicable, recognize and protect recreational, historical, biological, archaeological, and geological sites of special interest.	**Protect Special Sites Performance Measures:** To achieve and maintain certification, forest management practices, to the extent practicable, shall demonstrate concern for special sites.
Biodiversity Standard: Acceptable forest management includes the range of even and/or uneven age management practices.	**Biodiversity Performance Measures:** To achieve and maintain certification, members shall implement forest management practices that enhance the health and productivity of the woodland while considering biodiversity on a landscape or watershed scale.
Slash Disposal and Utilization Standard: Members shall consider harvest contract wording that addresses utilization and slash hazard reduction.	**Slash Disposal and Utilization Performance Measures:** To achieve and maintain certification, members shall make a good-faith effort to utilize, in an environmentally and/or economically sound manner, all severed and/or damaged materials on a harvest site.
Prudent Use of Chemicals Standard: Forest management practices using herbicides, pesticides, and/or fertilizers and implemented by the landowner shall be of the type that maintain or enhance the health and productivity of the woodland while protecting soil, water, fish, and wildlife resources.	**Prudent Use of Chemicals Performance Measures:** To achieve and maintain certification, the use of herbicides, pesticides, and/or fertilizers must meet or exceed all applicable label requirements as well as all local, state, and federal laws.

(Table continues)

Table 5.5   American Tree Farm System (ATFS) Standards, Guidelines, and Performance Measures (continued)

Standards and guidelines for forest landowners	Performance measures for Tree Farm certification
Forestry Contractor Use Standard: ATFS provides information, education, and assistance to forest landowners regarding forest management practices that will sustain or enhance forest productivity, wildlife habitat, water quality, and outdoor recreation.	**Forestry Contractor Use Performance Measures:** To achieve and maintain certification, members shall make a good-faith effort to ensure that loggers and contractors working on members' property are made aware of special requirements. In addition, members are encouraged to contract with loggers and other forest management contractors who have completed recommended training and education programs offered for their profession in their respective states and to ensure that such contractors are insured and comply with all state and federal regulations.

Source: American Tree Farm System. "About Tree Farming: Standards of Sustainability for Forest Certification Including Performance Measures and Field Indicators." Available: http://www.treefarmsystem.org/aboutfarming/standards.cfm; accessed April 18, 2003.

References

American Tree Farm System. "Standards of Sustainability for Forest Certification." http://www.treefarmsystem.org/aboutfarming/standards.cfm. Accessed on April 18, 2003.

Bailey, Robert G. 1995. "Description of the Ecoregions of the United States." USDA, U.S. Forest Service. http://www.nearctica.com/ecology/ecoreg/ecoprov.htm. Accessed on March 25, 2003.

———. 2003. http://www.fs.fed.us/land/ecosysmgmt/ecoreg1_home.html. Accessed on March 25, 2003.

Birch, Thomas W. 1994. *Private Forest-Land Owners of the United States, 1994.* Radnor, PA: USDA, U.S. Forest Service, Northeastern Forest Experiment Station.

Cubbage, F., J. O'Laughlin, and C. S. Bullock. 1993. *Forest Resource Policy.* New York: John Wiley & Sons.

Endgame Research Services: A Public Information Network. 2003. *Timberland Investment Management Organizations (TIMOs).* http://www.endgame.org/timo.html. Accessed on April 18, 2003.

Forest Stewardship Council. 2003. "Who We Are." http://www.fscoax. org/principal.htm. Accessed on April 18, 2003.

Hayward, J., and I. Vertinsky. 1999. "High Expectations, Unexpected Benefits: What Managers and Owners Think of Certification." *Journal of Forestry* 97 (2): 13–17.

Intertribal Timber Council. 1993. *An Assessment of Indian Forests and Forest Management in the United States.* November. The Indian Forest Management Assessment Team, 52pp. plus appendices.

MacCleery, D. W. 1996. *American Forests: A History of Resiliency and Recovery,* 4th revision. Durham, NC: Forest History Society.

The Montreal Process. 1998. "What Is the Montreal Process?" http:// www.mpci.org/whatis_e.html. Accessed on July 21, 2003.

Nogueron, R. 2002. *Low-Access Forests and Their Level of Protection in North America.* Washington, DC: World Resources Institute.

Pinchot Institute. 2003. "Facts about Non-Federal Forest Lands." http:// www.pinchot.org/pic/farmbill/Facts.html. Accessed on March 25, 2003.

Ricketts, T. H., E. Dinerstein, D. M. Olson, and C. J. Loucks. 1999. *Terrestrial Ecoregions of North America: A Conservation Assessment.* Covelo, CA: Island Press.

SmartWood. 2003. "Practical Conservation through Certified Forestry: Certification Guidelines and Applications." http://www.smartwood. org/guidelines/index.html. Accessed on April 18, 2003.

Souder, Jon, and Sally Fairfax. 1995. *State Trust Lands: History, Management, and Sustainable Use.* Lawrence: University of Kansas Press.

Sustainable Forestry Initiative. 2003. "About SFI." http://www.aboutsfi. org/about.asp. Accessed on April 18, 2003.

———. 2003. "Our Principles and Objectives." http://www.aboutsfi. org/about_principles.asp. Accessed on April 18, 2003.

U.S. Department of Agriculture (USDA), U.S. Forest Service. 2000. *2000 RPA Assessment of Forest and Range Lands.* http://www.fs.fed.us/ pl/rpa/rpaassess.pdf. Accessed on July 21, 2003.

———. 2001. *U.S. Forest Facts and Historical Trends.* http://fia.fs.fed.us/ library/ForestFactsMetric.pdf. Accessed on July 21, 2003.

———. 2003. http:// www.fs.fed.us. Accessed on July 21, 2003.

U.S. Department of the Interior, U.S. Bureau of Land Management. 2003. http://www.blm.gov. Accessed on February 21, 2003.

U.S. Department of the Interior, U.S. National Park Service. 2003. http://www.nps.gov. Accessed on February 21, 2003.

U.S. Department of the Interior, U.S. National Wilderness Preservation System. 2003. http://www.nwps.gov. Accessed on February 21, 2003.

World Resources Institute (WRI). 2002. *Low-Access Forests and Their Level of Protection in North America.* http://www.wri.org/wri/gfw/gfw_namerica.html. Accessed on March 25, 2003.

6

Directory of
U.S. Forestry Organizations

Numerous government and nongovernment organizations are involved in forest conservation, management, and utilization. In this chapter, the reader will find a whole host of organizations at state, national, and international levels chosen for their importance and contributions to U.S. forest conservation. The types of organizations listed include international and national government organizations, state government organizations, nongovernmental conservation organizations, forestry schools accredited by the Society of American Foresters, private forest products companies, forestry trade associations, forestry professional associations, and other forestry organizations that defy easy categorization. More information about any these organizations can be obtained by logging on to the web site address provided.

Federal and International
Government Organizations

Bureau of Indian Affairs
See **U.S. Department of the Interior, Bureau of Indian Affairs**

Bureau of Land Management
See **U.S. Department of the Interior, Bureau of Land Management**

Cooperative State Research, Education, and Extension Service
See U.S. Department of Agriculture, Cooperative State
Research, Education, and Extension Service

Council on Environmental Quality
722 Jackson Place NW
Washington, DC 20503
Phone: (202) 456-6224
Web site: www.whitehouse.gov/ceq/

The Council on Environmental Quality (CEQ) provides environmental expertise and policy analysis for the executive office of the
President. The council also implements statutory or regulatory requirements and programs related to the National Environmental
Policy Act of 1969, which applies to management of all Federal
public forest lands.

Environmental Protection Agency
See U.S. Environmental Protection Agency

Fish and Wildlife Service
See U.S. Department of the Interior, U.S. Fish and Wildlife
Service

Forest Service
See U.S. Department of Agriculture, U.S. Forest Service

Geological Survey
See U.S. Department of the Interior, U.S. Geological Survey

Ministry of Water, Lands, and Air Protection
P.O. Box 9339
Victoria, BC V8W 9M1
Canada
Phone: (604) 387-9422
Web site: http://www.gov.bc.ca/wlap/

The mission of the Ministry of Water, Land, and Air Protection is
to provide leadership in building environmental principles into
day-to-day decisions of governments, corporations, and private
individuals. Also, it monitors and reports on the state of the environment and ensures that defensible environmental standards are
set and complied with. The ministry manages natural habitats,

wildlife, and water resources for ecological diversity and economic and recreational opportunities.

National Oceanic and Atmospheric Administration
See **U.S. Department of Commerce, National Oceanic and Atmospheric Administration**

National Park Service
See **U.S. Department of the Interior, National Park Service**

National Science Foundation
4201 Wilson Boulevard
Arlington, VA 22230
Phone: (703) 306-1234
Web site: http://www.nsf.gov

The National Science Foundation supports science and engineering research and develops science education programs; it funds long-term research on forest ecosystems. Policy for the foundation is established through the National Science Board.

Natural Resource Conservation Service
See **U.S. Department of Agriculture, Natural Resource Conservation Service**

Natural Resources Canada
Canadian Forest Service
580 Booth Street
Ottawa, Ontario K1A 0E4
Canada
Web site: http://www.NRCan.gc.ca/cfs

Natural Resources Canada/Canadian Forest Service promotes sustainable development of forests and competitiveness of the forest sector. It also establishes links with nongovernmental organizations to address issues of international trade, market access, and worldwide sustainable forestry.

North American Wetlands Conservation Council
4401 North Fairfax Drive
Suite 110
Arlington, VA 22203
Phone: (703) 358-1784

E-mail: r9arw_nawwo@mail.fws.gov
Web site: http://www.fws.gov/r9nawwo/nawcahp.html

The North American Wetlands Conservation Council encourages public-private partnerships to conserve wetland ecosystems, including forested wetlands, for birds, fish, and wildlife.

St. Croix International Waterway Commission
Box 610
Calais, ME 04619
Phone: (506) 466-7550

The St. Croix International Waterway Commission is a joint commission of the state of Maine and the province of New Brunswick to implement international management plans for the St. Croix River system, which includes large amounts of forest land.

Tennessee Valley Authority
400 W. Summit Hill Drive
Knoxville, TN 37902-1499
Phone: (865) 632-2101
Web site: http://www.tva.gov/index.htm

The Tennessee Valley Authority (TVA) provides public service to the people of the Tennessee River valley by supporting sustainable economic development, supplying affordable and reliable power, and managing the Tennessee River system, which includes large amounts of forest land.

U.S. Department of Agriculture
Cooperative State Research, Education, and Extension Service
1400 Independence Avenue SW
Stop 2201
Washington, DC 20250-2201
Phone: (202) 720-7441
Web site: http://www.reeusda.gov/

The goal of the Cooperative Extension Service is to advance knowledge for agriculture, the environment, human health and well-being, and communities. This agency provides leadership to support university-based and other institutional research, education, and extension programs including forestry extension to private forest landowners.

U.S. Department of Agriculture
Natural Resource Conservation Service
USDA, NRCS, Office of the Chief
Fourteenth and Independence Avenue SW
Room 5105-A
Washington, DC 20250
Phone: (202) 720-7246
Web site: http://www.nrcs.usda.gov

The Natural Resources Conservation Service provides leadership in a partnership effort to help people conserve, maintain, and improve natural resources and environment.

U.S. Department of Agriculture
U.S. Forest Service
P.O. Box 96090
Washington, DC 20090-6090
Phone: (202) 205-8333

The Forest Service is a Federal agency that manages National Forests and National Grasslands. As the largest forestry research organization in the world, it provides technical and financial assistance to state and private forestry agencies. The mission of the Forest Service is to sustain the health, diversity, and productivity of the nation's forests and grasslands to meet the needs of current and future generations.

U.S. Department of Commerce
National Oceanic and Atmospheric Administration
Office of Global Program
1100 Wayne Avenue
Suite 1210
Silver Spring, MD 20910
Phone: (301) 427-2089
Web site: http://www.noaa.gov

The historical role of the National Oceanic and Atmospheric Administration (NOAA) has been to predict environmental changes, protect life and property, provide decision-makers with reliable scientific information, and foster global environmental stewardship. It provides coordination with national and international scientific communities for global warming, the Tropical Oceans and Global Atmosphere project, and climate research on issues including the interaction between forests and climate change.

U.S. Department of the Interior
Bureau of Indian Affairs
1849 C Street NW
Washington, DC 20240
Phone: (202) 208-5116
Web site: http://www.doi.gov/bureau-indian-affairs.html

The Bureau of Indian Affairs carries out the trust responsibility of the United States to Native American tribes. This includes the protection and enhancement of Native American lands and the conservation and development of forests/other natural resources.

U.S. Department of the Interior
Bureau of Land Management
Office of Public Affairs
1849 C Street
Room 406-LS
Washington, DC 20240
Phone: (202) 452-5125
Web site: http://www.blm.gov/

It is the mission of the Bureau of Land Management (BLM) to sustain the health, diversity, and productivity of the public lands for the use and enjoyment of present and future generations. The BLM administers 262 million acres of America's public land, much of it forested, primarily in 12 western states.

U.S. Department of the Interior
National Park Service
1849 C Street NW
Washington, DC 20240
Phone: (202) 208-6843
Web site: http://www.nps.gov

The National Park Service preserves unimpaired the natural and cultural resources and values of the National Park System for the enjoyment, education, and inspiration of this and future generations. It cooperates with partners to extend the benefits of natural and cultural resource conservation and outdoor recreation throughout the United States and the world.

U.S. Department of the Interior
U.S. Fish and Wildlife Service
1849 C Street NW

Room 3012
Washington, DC 20240
Phone: (202) 208-4717
Web site: http://www.fws.gov

The mission of the U.S. Fish and Wildlife Service is to administer a national network of lands and waters for the conservation, management, and, where appropriate, restoration of the fish, wildlife, and plant resources and their habitats within the United States for the benefit of current and future generations of Americans.

U.S. Department of the Interior
U.S. Geological Survey
USGS National Center
12201 Sunrise Valley Drive
Reston, VA 20192
Phone: (703) 648-4000
Web site: http://www.usgs.gov

The U.S. Geological Survey (USGS) serves the nation by providing reliable scientific information to describe and understand the Earth; minimize loss of life and property from natural disasters; manage forest, water, biological, energy, and mineral resources; and enhance and protect our quality of life.

U.S. Environmental Protection Agency
401 M Street SW
Washington, DC 20460
Phone: (202) 260-2090
Web site: http://epa.gov

The U.S. Environmental Protection Agency (EPA), housed in the executive branch of the U.S. government, strives to achieve systematic control and abatement of pollution through research, monitoring, standard setting, and enforcement. EPA works with the Forest Service and other Federal land management agencies to control and abate pollution in forested landscapes.

U.S. Forest Service
See **U.S. Department of Agriculture, U.S. Forest Service**

State Organizations

National Association of State Foresters
Hall of the States
444 North Capitol Street
Suite 540
Washington, DC 20001
Phone: (202) 624-5415
E-mail: nasf@sso.org
Web site: http://www.stateforesters.org/

The National Association of State Foresters provides management assistance and protection services for more than two-thirds of the nation's forests. It ensures that individual resource managers have the best technical, educational, and financial assistance available to help them achieve their objectives in an environmentally beneficial way. State forestry agencies also leverage state and local resources to develop urban and community forestry programs and to help protect all forests from wildfire, destructive pests, and diseases. The following is a list of individual associations by state:

Alabama
Timothy C. Boyce
Alabama Forestry Commission
513 Madison Avenue
Montgomery, AL 36130
Phone: (334) 240-9304
Fax: (334) 240-9390
E-mail: BoyceT@
    forestry.state.al.us

Alaska
Jeff Jahnke
Alaska Division of Forestry
State Forester's Office
550 West Seventh Avenue
Suite 1450
Anchorage, AK 99501
Phone: (907) 269-8474
Fax: (907) 269-8902
E-mail: Jeff_Jahnke@
    dnr.state.ak.us

American Samoa
Sheri S. Mann
Forestry Program Manager
P.O. Box 5319
ASCC/AHNR
Pago Pago, AS 96799
Phone: 011 (684) 699-1394
Fax: 011 (684) 699-5011
E-mail: ssuemann@yahoo.com

Arizona
Kirk Rowdabaugh
Arizona State Land
    Department
2901 W. Pinnacle Peak Road
Phoenix, AZ 85027-1002
Phone: (602) 255-4059
Fax: (602) 255-1781
E-mail: krowdabaughaz@
    cybertrails.com

Arkansas
John T. Shannon
Arkansas Forestry
    Commission
3821 West Roosevelt Road
Little Rock, AR 72204-6396
Phone: (501) 296-1941
Fax: (501) 296-1949
E-mail: john.shannon@mail.
    state.ar.us

California
Andrea E. Tuttle
Department of Forestry and
    Fire Protection
P.O. Box 944246
1416 Ninth Street
Room 1505
Sacramento, CA 94244-2460
Phone: (916) 653-7772
Fax: (916) 653-4171
E-mail: andrea_tuttle@
    fire.ca.gov

Colorado
James E. Hubbard
Colorado State Forest Service
Colorado State University
203 Forestry Building
Fort Collins, CO 80523
Phone: (970) 491-6303
Fax: (970) 491-7736
E-mail: jhubbard@lamar.
    colostate.edu

Connecticut
Donald H. Smith
Division of Forestry
79 Elm Street
Hartford, CT 06106
Phone: (860) 424-3630
Fax: (860) 424-4070

E-mail: don.smith@
    po.state.ct.us

Delaware
E. Austin Short III
Delaware Forest Service
2320 S. DuPont Highway
Dover, DE 19901
Phone: (302) 698-4548
Fax: (302) 697-6245
E-mail: AUSTIN@
    dda.state.de.us

District of Columbia
Mark Buscaino
Trees and Lands – District
    Government
1105 O Street SE
Washington, DC 20003
Phone: (202) 698-8361
Fax: (202) 724-1406
E-mail: mark.buscaino@dc.gov

Florida
L. Earl Peterson
Division of Forestry
3125 Conner Boulevard
Tallahassee, FL 32399-1650
Phone: (850) 488-4274
Fax: (850) 488-0863
E-mail: peterse@
    doacs.state.fl.us

Georgia
J. Fred Allen
Georgia Forestry Commission
P.O. Box 819
Macon, GA 31202-0819
Phone: (478) 751-3480
Fax: (478) 751-3465
E-mail: fallen@gfc.state.ga.us

Guam
David T. Limtiaco
Territorial Forester
Forestry and Soil Resources
   Division
192 Dairy Road
Mangilao, Guam 96923
Phone: (671) 735-3949
Fax: (671) 734-0111
E-mail: dlimti@mail.gov.gu

Hawaii
Michael G. Buck
Division of Forestry and
   Wildlife
1151 Punchbowl Street
Honolulu, HI 96813
Phone: (808) 587-0166
Fax: (808) 587-0160
E-mail: Michael_G_Buck@
   exec.state.hi.us

Idaho
Winston A. Wiggins
Idaho Department of Lands
954 W. Jefferson Street
P.O. Box 83720
Boise, ID 83720-0050
Phone: (208) 334-0200
Fax: (208) 334-2339
E-mail: wwiggins@
   idl.state.id.us

Illinois
Stewart Pequignot
Division of Forest Resources
2005 Round Barn Road
Champaign, IL 61821
Phone: (217) 278-5773
Fax: (217) 278-5763
E-mail: spequignot@dnrmail.
   state.il.us

Indiana
Burnell C. Fischer
Department of Natural
   Resources
402 W. Washington Street
Room W296
Indianapolis, IN 46204
Phone: (317) 232-4105
Fax: (317) 233-3863
E-mail: bfischer@
   dnr.state.in.us

Iowa
Mike Brandrup
Department of Natural
   Resources
Wallace Office Building
East Ninth and Grand Avenue
Des Moines, IA 50319
Phone: (515) 281-8657
Fax: (515) 281-6794
E-mail: Mike.Brandrup@
   dnr.state.ia.us

Kansas
Raymond G. Aslin
Kansas Forest Service
2610 Claflin Road
Manhattan, KS 66502-2798
Phone: (785) 532-3300
Fax: (785) 532-3305
E-mail: raslin@oznet.ksu.edu

Kentucky
Leah W. MacSwords
Kentucky Division of Forestry
627 Comanche Trail
Frankfort, KY 40601
Phone: (502) 564-4496
Fax: (502) 564-6553
E-mail: leah.macswords@
   mail.state.ky.us

Louisiana
Paul D. Frey
Office of Forestry
P.O. Box 1628
Baton Rouge, LA 70821
Phone: (225) 952-8002
Fax: (225) 922-1356
E-mail: Paul_F@ldaf.state.la.us

Maine
Thomas C. Doak
Maine Forest Service
22 State House Station
Harlow Building
Augusta, ME 04333
Phone: (207) 287-2791
Fax: (207) 287-8422
E-mail: tom.doak@state.me.us

Maryland
Steven W. Koehn
DNR – Forest Service
580 Taylor Avenue, E-1
Annapolis, MD 21401
Phone: (410) 260-8501
Fax: (410) 260-8595
E-mail: skoehn@
    dnr.state.md.us

Massachusetts
Warren E. Archey
Department of Environmental
    Management
P.O. Box 1433
Pittsfield, MA 01202
Phone: (413) 442-4963
Fax: (413) 442-5860
E-mail: Warren.Archey@
    state.ma.us

Michigan
Gerald J. Thiede

Michigan Department of
    Natural Resources
Forest Management Division
Mason Building, Eighth Floor
P.O. Box 30452
Lansing, MI 48909-7952
Phone: (517) 335-4225
Fax: (517) 373-2443
E-mail: thiedeg@state.mi.us

Minnesota
Michael Carroll
Division of Forestry
500 Lafayette Road
St. Paul, MN 55155-4044
Phone: (651) 296-4485
Fax: (651) 296-5954
E-mail: mike.carroll@
    dnr.state.mn.us

Mississippi
James L. Sledge Jr.
Mississippi Forestry
    Commission
301 N. Lamar Street
Suite 300
Jackson, MS 39201
Phone: (601) 359-1386
Fax: (601) 359-1349
E-mail: jsledge@
    mfc.state.ms.us

Missouri
Robert L. Krepps
Missouri Department of
    Conservation
P.O. Box 180
Jefferson City, MO 65102
Phone: (573) 751-4115 ext. 3300
Fax: (573) 526-6670
E-mail: kreppr@mail.
    conservation.state.mo.us

Montana
Donald K. Artley
DNRC – Forestry Division
2705 Spurgin Road
Missoula, MT 59804
Phone: (406) 542-4300
Fax: (406) 542-4217
E-mail: dartley@state.mt.us

Nebraska
Gary L. Hergenrader
Nebraska Forest Service
Plant Industry Building
Room 103
Lincoln, NE 68583-0815
Phone: (402) 472-2944
Fax: (402) 472-2964
E-mail: ghergenrader1@
    unl.edu

Nevada
Steve Robinson
Division of Forestry
1201 Johnson Street
Suite D
Carson City, NV 89706-3048
Phone: (775) 684-2512
Fax: (775) 687-4244
E-mail: forester@govmail.
    state.nv.us

New Hampshire
Philip Bryce
Division of Forests and Lands
P.O. Box 1856
172 Pembroke Road
Concord, NH 03302-1856
Phone: (603) 271-2214
Fax: (603) 271-6488
E-mail: p_bryce@dred.
    state.nh.us

New Jersey
James S. Barresi
State Forestry Service
P.O. Box 404
Trenton, NJ 08625-0404
Phone: (609) 292-2520
Fax: (609) 984-0378
E-mail: jbarresi@dep.state.nj.us

New Mexico
Toby A. Martinez
Forestry Division
P.O. Box 1948
Santa Fe, NM 87504-1948
Phone: (505) 476-3328
Fax: (505) 476-3330
E-mail: TAMartinez@
    State.nm.us

New York
Robert K. Davies
New York State Department of
    Environmental
    Conservation
625 Broadway
Albany, NY 12233-4250
Phone: (518) 402-9405
Fax: (518) 402-9028
E-mail: rkdavies@
    gw.dec.state.ny.us

North Carolina
Stanford M. Adams
North Carolina Division of
    Forest Resources
1616 Mail Service Center
Raleigh, NC 27699
Phone: (919) 733-2162 ext. 202
Fax: (919) 715-4350
E-mail: Stan.Adams@
    ncmail.net

North Dakota
Larry A. Kotchman
North Dakota Forest Service
307 First Street
Bottineau, ND 58318-1100
Phone: (701) 228-5422
Fax: (701) 228-5448
E-mail: Larry_Kotchman@
ndsu.nodak.edu

Ohio
Ronald G. Abraham
Division of Forestry
1855 Fountain Square Court,
H-1
Columbus, OH 43224
Phone: (614) 265-6690
Fax: (614) 447-9231
E-mail: Ron.Abraham@
dnr.state.oh.us

Oklahoma
Roger L. Davis
Oklahoma Department of
Agriculture
Forestry Services
P.O. Box 528804
Oklahoma City, OK 73152-3864
Phone: (405) 521-3864
Fax: (405) 522-4583
E-mail: rogerd@oda.state.ok.us

Oregon
Marvin Brown
Oregon Department of
Forestry
2600 State Street
Salem, OR 97310
Phone: (503) 945-7211
Fax: (503) 945-7212
E-mail: mbrown@odf.state.or.us

Pennsylvania
James R. Grace
Bureau of Forestry
P.O. Box 8552
Harrisburg, PA 17105-8552
Phone: (717) 787-2703
Fax: (717) 783-5109
E-mail: jagrace@state.pa.us

Puerto Rico
Diego Jimenez Carrion
Forest Service Bureau – DNER
P.O. Box 9066600
Puerta de Tierra
San Juan, PR 00906-6600
Phone: (787) 724-3647
Fax: (787) 721-5984
E-mail: prforests@hotmail.com

Rhode Island
Thomas A. Dupree
Division of Forest
Environment
1037 Hartford Pike
North Scituate, RI 02857
Phone: (401) 647-3367
Fax: (401) 647-3590
E-mail: riforestry@edgenet.net

South Carolina
Bob Schowalter
South Carolina Forestry
Commission
P.O. Box 21707
Columbia, SC 29221
Phone: (803) 896-8800
Fax: (803) 798-8097
E-mail: bschowalter@
forestry.state.sc.us

South Dakota
Ray Sowers
Resource Conservation and
  Forestry
Foss Building
523 East Capitol Avenue
Pierre, SD 57501
Phone: (605) 773-3623
Fax: (605) 773-4003
E-mail: Ray.Sowers@
  state.sd.us

Tennessee
Director
Tennessee Department of
  Agriculture
Division of Forestry
P.O. Box 40627, Melrose
  Station
Nashville, TN 37204
Phone: (615) 837-5411
Fax: (615) 837-5003
E-mail: lburrow@
  mail.state.tn.us

Texas
James B. Hull
Texas Forest Service
301 Tarrow Drive
Suite 364
College Station, TX 77840-7896
Phone: (979) 458-6600
Fax: (979) 458-6610
E-mail: jim-hull@tamu.edu

Utah
A. Joel Frandsen, Director
Department of Natural
  Resources
1594 West North Temple
Suite 3520
Salt Lake City, UT 84114-5703
Phone: (801) 538-5540

Fax: (801) 533-4111
E-mail: joelfrandsen@utah.gov

Vermont
Conrad M. Motyka
Department of Forests, Parks
  and Recreation
103 S. Main Street
Waterbury, VT 05671-0601
Phone: (802) 241-3670
Fax: (802) 244-1481
E-mail: cmotyka@
  fpr.anr.state.vt.us

Virginia
James W. Garner
Virginia Department of
  Forestry
900 Natural Resources Drive
Suite 800
Charlottesville, VA 22903
Phone: (434) 977-6555
Fax: (434) 977-7749
E-mail: garnerj@dof.state.va.us

Washington
Pat McElroy
Department of Natural
  Resources
Box 47037
1111 Washington Street
Olympia, WA 98504-7037
Phone: (360) 902-1603
Fax: (360) 902-1775
E-mail: pat.mcelroy@
  wadnr.gov

West Virginia
C. Randall Dye
West Virginia Forestry
  Division
1900 Kanawha Boulevard East
Charleston, WV 25305-0180

Phone: (304) 558-3446
Fax: (304) 558-0143
E-mail: RDye@
   gwmail.state.wv.us

Wisconsin
Paul DeLong
DNR – Division of Forestry
P.O. Box 7921
Madison, WI 53707
Phone: (608) 266-2694
Fax: (608) 266-6983
E-mail: delong@dnr.state.wi.us

Wyoming
Tom Ostermann
Wyoming State Forestry
   Division
1100 West Twenty-Second
   Street
Cheyenne, WY 82002
Phone: (307) 777-7586
Fax: (307) 777-5986
E-mail: toster@state.wy.us

Nongovernment Conservation Organizations

Adopt-A-Stream Foundation
600 128th Street SE
Everett, WA 98208
Phone: (425) 316-8592
E-mail: aasf@streamkeeper.org
Web site: http://www.streamkeeper.org

The mission of the Adopt-A-Stream Foundation is to empower people to become stewards of watersheds, wetlands, and streams, many of which run through forests. The foundation's long-term goal is to ensure that all streams are adopted by local residents.

Aldo Leopold Foundation Inc.
E12919 Levee Road
P.O. Box 77
Baraboo, WI 53919
Phone: (608) 355-0279
E-mail: mail@aldoleopold.org
Web site: http://www.aldoleopold.org

The Aldo Leopold Foundation promotes harmony between people and land including forest land through restoration, research, and educational activities.

American Conservation Association Inc.
1200 New York Avenue NW
Suite 400
Washington, DC 20005
Phone: (202) 289-2431

The American Conservation Association is an educational and scientific organization formed to advance knowledge and understanding of conservation and to preserve and develop natural resources for public use.

American Forests
P.O. Box 2000
Washington, DC 20013
Phone: (202) 955-4500
E-mail: info@amfor.org
Web site: http://www.americanforests.org

American Forests plants trees for environmental restoration, is a pioneer in the science and practice of urban forestry, and communicates the benefits of trees and forests. Its focus is on assisting communities in planning and implementing tree and forest actions to restore and maintain healthy ecosystems and communities. It also works with community-based forestry partners in both urban and rural areas to help them participate in National Forest policy discussions.

American Hiking Society
1422 Fenwick Lane
Silver Spring, MD 20910
Phone: (301) 565-6704
E-mail: info@americanhiking.org
Web site: http://www.americanhiking.org

The American Hiking Society is a recreation-based conservation group dedicated to establishing, protecting, and maintaining foot trails, many of which run through forest land. The AHS lobbies to encourage funding for trails and promotes volunteerism in trail building and maintenance.

American Lands
726 Seventh Street SE
Washington, DC 20003
Phone: (202) 547-9400

E-mail: wafcdc@americanlands.org
Web site: http://www.americanlands.org

The mission of American Lands is to protect and recover North American native forests, grassland, and aquatic ecosystems and preserve biological diversity. American Lands strengthens grassroots conservation networks, provides advocacy services, and improves coordination and communication among groups.

Ancient Forest International
P.O. Box 1850
Redway, CA 95560
Phone: (707) 923-3015
E-mail: afi@ancientforest.org
Web site: http://www.ancientforest.org

Ancient Forest International is an alliance of conservationists dedicated to helping preserve, study, and increase awareness of the Earth's intact forest ecosystems while providing habitat continuity through corridors. Currently work is being done in Chile, Mexico, Ecuador, and the north Pacific Coast.

Biomass Users Network
383 Franklin Street
Bloomfield, NJ 07003
Phone: (201) 680-9100

The Biomass Users Network works to advance rural economic development in Third World countries in an environmentally sound manner through the innovative production and efficient use of biomass resources including forest resources.

Canada–United States Environmental Council
1101 Fourteenth Street NW
Suite 1400
Washington, DC 20005
Phone: (202) 682-9400
Web site: http://www.defenders.org

The Canada–United States Environmental Council is sponsored by Canadian and American conservation and environmental groups to facilitate interchange of information and cooperative action on environmental issues of concern including forest conservation issues.

Canadian Parks and Wilderness Society
880 Wellington Street, Suite 506
Ottawa, Ontario K1R 6K7
Canada
Phone: (613) 569-7226
E-mail: info@cpaws.org
Web site: http://www.cpaws.org

The Canadian Parks and Wilderness Society is an advocacy organization dedicated to the protection of Wilderness areas and the preservation and stewardship of Canada's public lands.

Canadian Wildlife Federation
2740 Queensview Drive
Ottawa, Ontario K2B 1A2
Canada
Phone: (613) 721-2286

The Canadian Wildlife Federation fosters understanding of natural processes so that people may live in harmony with the land and its resources, including forests, for the long-term benefit of society. It also maintains educational programs and sponsors research.

Center for Watershed Protection
8391 Main Street
Ellicott, MD 21043
Phone: (410) 461-8323
E-mail: center@cwp.org
Web site: http://www.cwp.org

The Center for Watershed Protection is dedicated to new, cooperative ways of protecting and restoring watersheds such as forest conservation.

Coastal Alliance
600 Pennsylvania Avenue SE
Suite 340
Washington, DC 20003
Phone: (202) 546-9554
E-mail: coast@coastalliance.org
Web site: http://www.coastalliance.org

The Coastal Alliance is a public interest group dedicated to raising public awareness about coastal resources such as coastal forests. It provides information about activities affecting the nation's four coasts.

Coastal Society
P.O. Box 25408
Alexandria, VA 22313
Phone: (703) 768-1599
E-mail: coastal@aol.com
Web site: http://www.coastalsociety.org

The Coastal Society is an organization of private, academic, and governmental professionals and students dedicated to addressing coastal issues such as coastal forest conservation, fostering dialog, forging partnerships, and promoting communication and education.

Conservation International
1919 M Street NW
Suite 600
Washington, DC 20036
Phone: (202) 429-5660
Web site: http://www.conservation.org

Conservation International (CI) is dedicated to the preservation of tropical and temperate ecosystems. CI works in partnerships with indigenous peoples and with organizations to sustain biological diversity and the ecological processes that support life on Earth.

Conservation Law Foundation
62 Summer Street
Boston, MA 02110
Phone: (617) 350-0990
Web site: http://www.clf.org

The Conservation Law Foundation is an environmental law organization dedicated to improving resource management, environmental protection, and public health in New England.

Defenders of Wildlife
National Headquarters
1101 Fourteenth Street NW
Suite 1400
Washington, DC 20005

Phone: (202) 682-9400
E-mail: info@defenders.org
Web site: http://www.defenders.org

Defenders of Wildlife is dedicated to the protection of all native wild animals and plants in their natural communities. It focuses its programs on the accelerating rate of extinction of species and the associated loss of biological diversity and on habitat alteration and destruction. Its programs encourage protection of entire ecosystems and interconnected habitats while protecting predators that serve as indicator species for ecosystem health.

Earthstewards Network
P.O. Box 10697
Bainbridge Island, WA 98110
Phone: (206) 842-7986
E-mail: office@earthstewards.org
Web site: http://www.earthstewards.org

Earthstewards Network is the international network for global conflict resolution. Rainforest reforestation and urban forestry projects are used to bring together peoples of cultures in conflict to improve environmental conditions.

Earthwatch Institute
3 Clocktower Place, Suite 100
P.O. Box 75
Maynard, MA 01754
Phone: (978) 461-0081
E-mail: info@earthwatch.org
Web site: http://www.earthwatch.org

Earthwatch Institute sponsors scientific field research worldwide. It recruits volunteers to help field scientists including forest scientists with their research in 50 countries.

Eastern Shore Land Conservancy
P.O. Box 169
Queenstown, MD 21658
Phone: (410) 827-9756
E-mail: eslcmain@usa.net
Web site: http://www.eslc.org

The Eastern Shore Land Conservancy preserves farms, forests, and natural areas for future generations using voluntary land-protection tools.

Environmental Defense
257 Park Avenue South
New York, NY 10010
Phone: (212) 505-2100
E-mail: members@environmentaldefense.org
Web site: http://www.environmentaldefense.org

Environmental Defense is dedicated to protecting the environmental rights of all people, including future generations. Among these rights are clean air and water, healthy and nourishing food, and a flourishing ecosystem. It evaluates environmental problems and works to create and advocate solutions that win lasting political, economic, and social support because they are nonpartisan, cost-efficient, and fair.

Environmental Law and Policy Center of the Midwest
25 East Wacker Drive
Suite 1300
Chicago, IL 60601
Phone: (312) 673-6500
E-mail: elpc@elpc.org
Web site: http://www.elpc.org

The Environmental Law and Policy Center of the Midwest is a public interest environmental advocacy organization working to implement sustainable energy strategies, promote innovative transportation approaches, expand and develop green markets, and develop sound environment and forest management practices in the Midwest.

Europarc Federation
Kroellstrasse 5
D-94481 Grafenau
Grafenau 94481
Germany
Phone: 4985-529-6100
E-mail: office@europarc.org
Web site: http://www.europarc.org

The Europarc Federation promotes and supports protected areas in Europe. It aims to facilitate the exchange of technical and scientific expertise, information, and personnel between parks and reserves. It organizes training and exchange programs and provides professional advice on the establishment and development of protected areas.

Forest Service Employees for Environmental Ethics
P.O. Box 11615
Eugene, OR 97440
Phone: (541) 484-2692
E-mail: fseee@fseee.org
Web site: http://www.fseee.org

Forest Service Employees for Environmental Ethics is a national organization of Forest Service employees and retirees, other resource professionals, and concerned citizens working to change from within the Forest Service's basic management philosophy to a land ethic that ensures ecologically and economically sustainable management.

Forest Trust
P.O. Box 519
Santa Fe, NM 87504
Phone: (505) 983-8992
E-mail: forest@theforesttrust.org
Web site: http://www.theforesttrust.org

The Forest Trust is dedicated to protecting the integrity of the forest ecosystem and improving the lives of people in rural communities. It provides protection strategies to grassroots environmental organizations, rural communities, and public agencies. It also provides land management services to private landowners.

Friends of the Earth
1025 Vermont Avenue NW
Washington, DC 20005-6303
Phone: (877) 843-8687
E-mail: foe@foe.org
Web site: http://www.foe.org

A national, nonprofit advocacy organization, Friends of the Earth is dedicated to protecting the planet from environmental degradation; preserving biological, cultural, and ethnic diversity; and empowering citizens to have an influential voice in discussions affecting the quality of their environment and their lives.

Grand Canyon Trust
HC 64, Box 1801
Moab, UT 84532
Phone: (435) 259-5284
Web site: http://www.grandcanyontrust.org

The Grand Canyon Trust is a regional organization dedicated to protecting the Colorado Plateau, which contains abundant forest land.

Grassland Heritage Foundation
P.O. Box 394
Shawnee Mission, KS 66201
Phone: (913) 262-3506
E-mail: grasslandheritage@grapevine.net
Web site: http://www.grasslandheritage.org

The Grassland Heritage Foundation is dedicated to prairie preservation and education. It encourages the preservation of all remaining prairies and works to increase public awareness of our prairie heritage. Many National Grasslands are managed by the U.S. Forest Service. In many prairie ecosystems, prairie and forests are interspersed.

Greater Yellowstone Coalition
P.O. Box 1874
13 S. Wilson, Suite 2
Bozeman, MT 59771
Phone: (406) 586-1593
E-mail: gyc@greateryellowstone.org
Web site: http://www.greateryellowstone.org

The Greater Yellowstone Coalition is dedicated to preserving and protecting the greater Yellowstone ecosystem by enhancing the ecosystem concept, raising the national public consciousness about the area, and combining the political effectiveness of the coalition's members.

Heartwood
P.O. Box 1424
Bloomington, IN 47402
Phone: (812) 337-8898
E-mail: info@heartwood.org
Web site: http://www.heartwood.org

Heartwood works to protect the forest by addressing responsible and sustainable stewardship of privately held forestland; the influx of high-volume, indiscriminate forest-consuming industries; and the need to dramatically reduce the wasteful use of wood and paper that drives the destruction of the forest.

International Ecology Society
1471 Barclay Street
St. Paul, MN 55106
Phone: (612) 579-7008

The International Ecology Society is a volunteer-staffed organization dedicated to the protection of the environment and the encouragement of better understanding of all life forms.

International Society for the Preservation of the Tropical Rainforests
3931 Camino de la Cumbre
Sherman Oaks, CA 91423
Phone: (818) 788-2202
E-mail: forest@nwc.net

The International Society for the Preservation of the Tropical Rainforests is dedicated to the global conservation of tropical forest resources through the promotion of park implementation, sustainable agriculture, and timber harvesting.

International Union for Conservation of Nature and Natural Resources
Rue Mauverney 28, CH-1196
Gland, Switzerland
Phone: 022-999-0001
Web site: http://www.lucn.org

The International Union for Conservation of Nature and Natural Resources is an independent body the goal of which is to promote scientifically based action for the conservation of nature and to ensure that development is environmentally and socially sustainable. It consists of 880 members and maintains a global network of more than 6,000 scientists.

Island Resources Foundation
1718 P Street NW
Suite T-4

Washington, DC 20036
Phone: (202) 265-9712
E-mail: irf@irf.org
Web site: http://www.irf.org

Island Resources Foundation, an independent center for the study of island systems, is dedicated to improved resources management; comprehensive development planning; and the conservation of cultural, physical, and natural resources of islands.

Izaak Walton League of America Inc.
707 Conservation Lane
Gaithersburg, MD 20878
Phone: (301) 548-0150
Web site: http://www.iwla.org

The Izaak Walton League of America promotes means and opportunities for educating the public to conserve, maintain, protect, and restore the soil, forest, water, air, and other natural resources. It also promotes the enjoyment and wholesome use of those resources.

Mid-Atlantic Council of Watershed Associations
12 Morris Road
Ambler, PA 19002

The Mid-Atlantic Council of Watershed Associations promotes the exchange of ideas about citizen watershed association activities and advises any group wishing to start a new watershed association. Most watersheds in the Mid-Atlantic contain abundant forest land.

National Arbor Day Foundation
100 Arbor Avenue
Nebraska City, NE 68410
Phone: (402) 474-5655
Web site: http://www.arborday.org

The National Arbor Day Foundation sponsors Trees for America, Arbor Day, Tree City USA, Conservation Trees, and Rainforest Rescue educational programs.

National Association of Recreation Resource Planners
P.O. Box 2430

Pensacola, FL 32513

The National Association of Recreation Resource Planners exchanges recreation resource planning information among Federal, state, and regional agencies. It participates in national recreation concerns, promotes improvements in the state of the art of recreation planning, and advocates for conservation and recreation opportunities. Its headquarters, now in Pensacola, rotate yearly.

National Audubon Society
700 Broadway
New York, NY 10003
Phone: (212) 979-3000
Web site: http://www.audubon.org

The National Audubon Society has dedicated itself to bringing about a national and worldwide culture of conservation by focusing on the conservation of birds and other wildlife and their habitats, expanding educational programs to nurture appreciation of nature and understanding of the essential link between ecological health and the well-being of human civilization, and investing heavily in Audubon's distinctive grassroots network as the primary instrument of its environmental advocacy.

National Conservation Foundation
509 Capitol Court NE
Washington, DC 20002-4946
Phone: (202) 547-6223
Web site: http://www.nacdnet.org

Directed by the Conservation Districts Foundation Inc., the National Conservation Foundation collects conservation and environmental education materials about all natural resources, including forests.

National Forest Foundation
1050 Seventeenth Street NW
Suite 600
Washington, DC 20036
Phone: (202) 496-4963

The National Forest Foundation seeks to build relationships that result in measurable improvements in the health, productivity, and diversity of national forests and grasslands.

National Forestry Association
374 Maple Avenue E
Suite 310
Vienna, VA 22180
Phone: (703) 255-2300
Web site: http://www.forestry.tv

The National Forestry Association is a nationwide advocate of sustainable forestry on private and public lands. Programs include the National Forestry Network, Green Tag Forestry, the American Hardwood Management Advisory Board, and the National Historic Lookout Register.

National Network of Forest Practitioners
P.O. Box 487
Santa Fe, NM 87504
Phone: (505) 995-0000
E-mail: info@nnfp.org
Web site: http://www.nnfp.org

The National Network of Forest Practitioners is a grassroots alliance of rural people, organizations, and businesses finding practical ways to integrate economic development, environmental protection, and social justice.

National Park and Conservation Association
1300 Nineteenth Street NW
Suite 300
Washington, DC 20036
Phone: (202) 223-6722
E-mail: natparks@aol.com
Web site: http://www.npca.org

The National Park and Conservation Association, a citizen organization, is dedicated to preserving, protecting, and enhancing the national park system. It has been an advocate and critic of the National Park Service.

National Park Trust
415 Second Street NE
Suite 210
Washington, DC 20002
Phone: (202) 548-0500
E-mail: nptrust@aol.com

Web site: http://www.parktrust.org

The National Park Trust is dedicated to protecting resources within and around parklands and other natural and historic properties. It is the only private citizen group recognized by Congress to own and manage, in cooperation with the National Park Service, a unit of public land.

National Wildlife Federation
11100 Wildlife Center Drive
Reston, VA 20190
Phone: (703) 438-6000
Web site: http://www.nwf.org

The mission of the National Wildlife Federation is to educate, inspire, and assist individuals and organizations of diverse cultures to conserve wildlife and other natural resources and to protect the Earth's environment to achieve a peaceful, equitable, and sustainable future.

Native Ecosystems Council
P.O. Box 125
Willow Creek, MT 59760

Native Ecosystems Council is a nonprofit technical assistance group doing work on all the forests in the Rocky Mountain region. With special expertise in forest planning policy, it is currently submitting management alternatives based on principles of conservation biology and landscape ecology.

Native Forest Network
P.O. Box 8251
Missoula, MT 59807
Phone: (406) 251-2385
E-mail: nfn@wildrockies.org
Web site: http://nativeforest.org

The Native Forest Network is a global autonomous collective of forest activists, indigenous peoples, conservation biologists, and nongovernmental organizations. It functions on a consensus basis and is nonviolent, nonhierarchical, and nonpatriarchal. Its mission is to protect the world's remaining native forests, be they temperate or otherwise, to ensure they can survive, flourish, and maintain their evolutionary potential.

Natural Resources Council of America
1025 Thomas Jefferson Street NW
Suite 109
Washington, DC 20007
Phone: (202) 333-0411
E-mail: nrca@nrcanet.org

The Natural Resources Council of America is an association of nonprofit environmental and conservation organizations dedicated to the protection, conservation, and responsible management of the nation's natural resources.

Natural Resources Defense Council
40 West Twentieth Street
New York, NY 10011
Phone: (212) 727-2700
E-mail: nrdcinfo@nrdc.org
Web site: http://www.nrdc.org

The Natural Resources Defense Council combines legal action, scientific research, and citizen education in programs to protect natural resources and improve the quality of the human environment. It publishes a quarterly magazine, *On Earth: Environment, Politics, and People,* and regularly reports on a variety of environmental issues.

Nature Conservancy
4245 North Fairfax Drive
Suite 100
Arlington, VA 22203-1606
Phone: (800) 628-6860
E-mail: comment@tnc.org
Web site: http://www.nature.org

The goal of the Nature Conservancy is to preserve the plants, animals, and natural communities that represent the diversity of life on Earth by protecting the lands and waters they need to survive. It works closely with communities, businesses, and people, taking a nonconfrontational approach and using sound science.

Nature Conservancy of Canada
110 Eglinton Avenue W
Suite 400
Toronto, Ontario M4R 1A3

Canada
Phone: (416) 932-3202
E-mail: nature@natureconservancy.ca
Web site: http://www.natureconservancy.ca

The Nature Conservancy of Canada is the only national charity dedicated to preserving ecologically significant areas and places of special beauty and educational interest through outright purchase, donations, and conservation agreements.

Nature Conservation Society of Japan
Nihon-Shizen-Hogo-Kyokai
Yamaji Saubauncho Building 3F
5-24 Sanbancho
Chiyoda Ku, Tokyo 102
Japan
Phone: 8133-265-0521

The Nature Conservation Society of Japan is a conservation organization devoted to promoting conservation, research, and education concerning the natural areas and wildlife in Japan.

North Cascades Conservation Council
P.O. Box 95980
Seattle, WA 98145
Phone: (206) 282-1644

The North Cascades Conservation Council seeks to protect and preserve the North Cascades' scenic, scientific, recreational, educational, wildlife, and wilderness values from the Columbia River to the U.S.-Canadian border.

Northcoast Environmental Center
879 Ninth Street
Arcata, CA 95521
Phone: (707) 822-6918
E-mail: nec@igc.org
Web site: http://www.necandeconews.to

The Northcoast Environmental Center, an educational organization, is dedicated to illuminating the relationships between humankind and the biosphere.

Northwest Ecosystem Alliance
1421 Cornwall Avenue

Suite 201
Bellingham, WA 98225
Phone: (360) 671-9950
E-mail: nwea@ecosystem.org
Web site: http://www.ecosystem.org

The Northwest Ecosystem Alliance protects and restores wildlands in the Pacific Northwest and supports such efforts in British Columbia. The alliance bridges sciences and works with activists, policymakers, and the public to conserve our natural heritage.

Ozarks Resource Center
P.O. Box 1198
Ava, MO 65608
Phone: (417) 683-6245
E-mail: jlorrain@goin.missouri.org

The center provides research, education, technical assistance, and dissemination of information on renewable resources–based technology; sustainable agriculture; environmentally responsible practices; sustainable community economic development; and self-reliance for the family, farm, community, and Ozarks and other bioregions.

Pacific Rivers Council
P.O. Box 10798
Eugene, OR 97440
Phone: (541) 345-0119
E-mail: pacificriver@igc.apc.org

The council protects and restores rivers, their watersheds, and native aquatic species.

Partners in Parks
4916 Butterworth Place NW
Washington, DC 20016
Phone: (202) 364-7244
E-mail: partpark@cqi.com

This organization encourages, promotes, and establishes professional-level partnerships between National Park and other public land managers and those who would contribute their time and skills to studying, protecting, and interpreting natural and cultural features.

Pinchot Institute for Conservation
1616 P Street NW
Suite 100
Washington, DC 20036
Phone: (202) 797-6580
E-mail: pinchot@pinchot.org
Web site: http://www.pinchot.org

The Pinchot Institute for Conservation is a nonprofit natural resource policy, research, and education organization dedicated to
leadership in conservation thought, policy, and action. The institute was designed to facilitate communication and closer cooperation among resource managers, scientists, policymakers, and the
American public.

Public Employees for Environmental Responsibility
2001 S Street NW
Suite 570
Washington, DC 20009
Phone: (202) 265-7337
E-mail: info@peer.org
Web site: http://www.info@peer.org

Public Employees for Environmental Responsibility (PEER) is an
alliance of land managers, scientists, biologists, law enforcement
officials, and other government professionals including many
Forest Service employees dedicated to the protection of the nation's environment. PEER advocates the responsible management
of natural resources and promotes environmental ethics, professional integrity, and accountability.

Public Lands Foundation
P.O. Box 7226
Arlington, VA 22207
Phone: (703) 790-1988
E-mail: leaplf@erols.com
Web site: http://www.publicland.org

Administered by the Bureau of Land Management, the foundation is an independent advocate to keep the public lands public
and to properly use and protect the public land.

Rainforest Action Network
221 Pine Street

Suite 500
San Francisco, CA 94104
Phone: (415) 398-4404
E-mail: rainforest@ran.org
Web site: http://www.ran.org

Rainforest Action Network (RAN) works nationally and internationally on major campaigns using nonviolent direct action to protect rainforests and defend the rights of indigenous people. RAN also produces educational materials, teachers' packets, and fact sheets.

Rainforest Relief
P.O. Box 150566
Brooklyn, NY 11215
Phone: (718) 398-3760
E-mail: relief@igc.org
Web site: http://www.enviroweb.org/rainrelief

Rainforest Relief works through education and nonviolent direct action to end the loss of tropical and temperate rainforests by reducing the demand for products and materials for which rainforests are destroyed.

Renewable Natural Resources Foundation
5430 Grosvenor Lane
Bethesda, MD 20814
Phone: (301) 493-9101
E-mail: infor@rnrf.org
Web site: http://www.rnrf.org

The foundation conducts conferences and symposia about renewable natural resource subjects and public policy alternatives.

Resources for the Future
1616 P Street NW
Washington, DC 20036-1400
Phone: (202) 328-5000
Web site: http://www.rff.org

RFF is a nonprofit and nonpartisan think tank that conducts independent research, rooted primarily in economics and other social sciences, on environmental and natural resource issues. RFF researchers work on a variety of issues ranging from climate change to electric utility restructuring to sustainable forestry.

Save America's Forests
4 Library Court SE
Washington, DC 20003
Phone: (202) 544-9219
Web site: http://www.saveamericasforests.org

This nationwide coalition of grassroots regional and national environmental groups, public interest groups, and responsible businesses and individuals works to pass strong forest protection legislation in Congress.

Sierra Club
85 Second Street, Second Floor
San Francisco, CA 94105
Phone: (415) 977-5500
E-mail: information@sierraclub.org
Web site: http://www.sierraclub.org

The mission of the Sierra Club is to explore, enjoy, and protect the wild places of the Earth; practice and promote the responsible use of the Earth's ecosystems and resources; educate and enlist humanity to protect and restore the quality of the natural and human environment; and use all lawful means to carry out these objectives.

Society for Ecological Restoration
1207 Seminole Highway
Suite B
Madison, WI 53711
Phone: (608) 262-9547

The society was created to promote the development of ecological restoration, both as a discipline and as a model for a healthy relationship with nature, and to raise awareness of the value and limitations of restoration as a conservation strategy.

Southwest Center for Biological Diversity
P.O. Box 710
Tucson, AZ 85702-0710
Phone: (520) 623-5252
E-mail: center@biologicaldiversity.org
Web site: http://www.sw-center.org

Combining conservation biology with litigation, policy advocacy, and an innovative strategic vision, the Center for Biological Diversity is working to secure a future for animals and plants hovering on the brink of extinction, for the wilderness they need to survive, and by extension for the spiritual welfare of generations to come.

Theodore Roosevelt Conservation Alliance
2409 Dearborn
Suite K
Missoula, MT 59801
Phone: (406) 541-9975
E-mail: info@trca.org
Web site: http://www.trca.org

The alliance's mission is to inform and engage Americans to foster the nation's conservation legacy while working to nurture, enhance, and protect fish, wildlife, and habitat resources on public lands.

Trust for Public Land
116 New Montgomery Street
Fourth Floor
San Francisco, CA 94105
Phone: (415) 495-4014
E-mail: info@tpl.org
Web site: http://www.tpl.org

The Trust for Public Land helps local communities, public agencies, and other nonprofits acquire natural, scenic, and recreational land for permanent protection.

UN Environment Programme
P.O. Box 30552
Nairobi, Kenya
Phone: 254-262-3089
E-mail: ipainfo@unep.org
Web site: http://www.unep.org

The UNEP was established by the UN General Assembly to be the environmental conscience of the United Nations system. It assesses the state of the world's environment; acts in an environmental management capacity for developing countries; and raises environmental considerations for the social and economic policies and programs of UN agencies.

Union of Concerned Scientists
2 Brattle Square
Cambridge, MA 02238
Phone: (617) 547-5552
E-mail: ucs@ucsusa.org
Web site: http://www.ucsusa.org

UCS is an independent nonprofit alliance of 50,000 concerned citizens and scientists across the country. It augments rigorous scientific analysis with innovative thinking and committed citizen advocacy to build a cleaner, healthier environment and a safer world.

U.S. Public Interest Research Group
218 D Street SE
Washington, DC 20003
Phone: (202) 546-9707
E-mail: uspirg@pirg.org
Web site: http://www.pirg.org

U.S. Public Interest Research Group (PIRG) is the national lobbying office for state PIRGs around the country, representing more than a million members. Its mission is to conduct independent research and to lobby for national environmental and consumer protections.

Wilderness Society
900 Seventeenth Street NW
Washington, DC 20006
Phone: (202) 833-2300
E-mail: member@tws.org
Web site: http://www.wilderness.org

The Wilderness Society works to protect America's wilderness and to develop a nationwide network of wildlands through public education, scientific analysis, and advocacy. Its goal is to ensure that future generations will enjoy the clean air and water, wildlife, beauty, and opportunities for recreation and renewal that pristine forests, rivers, deserts, and mountains provide.

Wildlife Action Inc.
P.O. Box 866
Mullins, SC 29574

Phone: (843) 464-8473
Web site: http://www.wildlifeaction.com

Wildlife Action is dedicated to the appreciation and enjoyment of our wildlife heritage and to educating the public in the value of protection, restoration, enhancement, and wise use of our natural resources.

Women's Environment and Development Organization
355 Lexington Avenue, Third Floor
New York, NY 10017
Phone: (212) 973-0325
E-mail: wedo@igc.apc.org
Web site: http://www.wedo.org

This is an international advocacy network actively working to transform society to achieve social, political, economic, and environmental justice for all through the empowerment of women. Many projects focus on increasing women's access and role in sustainable forest management.

World Forestry Center
4033 S.W. Canyon Road
Portland, OR 97221
Phone: (503) 228-1367
Web site: http://www.worldforestry.org

The center promotes a greater appreciation and understanding of the world's forests and related natural resources. It operates a forestry museum, an arboretum, and an 80-acre demonstration forest.

World Parks Endowment Inc.
1616 P Street NW
Suite 200
Washington, DC 20036
Phone: (202) 797-6540
E-mail: worldparks@juno.com
Web site: http://www.worlparks.org

This international organization acquires land in the rainforest and other critical sites for biological diversity. It provides funds for park management of tropical rainforests and other ecosystems of great conservation importance.

World Resources Institute
10 G Street NE
Suite 800
Washington, DC 20002
Phone: (202) 729-7600
Web site: http://www.wri.org

WRI is an environmental think tank that goes beyond research to find practical ways to protect the Earth and improve people's lives. It provides information, ideas, and solutions to global environmental problems.

World Wildlife Fund, U.S.
1250 Twenty-Fourth Street NW
Washington, DC 20037-1175
Phone: (202) 293-4800
Web site: http://www.panda.org

WWF uses a factual, science-based approach to conservation that focuses on several global priorities and recognizes the need to strike a balance with other economic and social considerations to achieve concrete results. WWF's work focuses on species conservation; the protection of forest, freshwater, ocean, and coastal ecosystems; and the fight against climate change and toxic pollution.

Worldwatch Institute
1776 Massachusetts Avenue NW
Washington, DC 20036
Phone: (202) 452-1999
E-mail: worldwatch@worldwatch.org
Web site: http://www.worldwatch.org

This research institute informs policymakers and the public about emerging global problems and trends and the complex links between the world economy and its environmental support systems.

Society of American Foresters Accredited Forestry Schools

Professional accreditation of an academic program defines the knowledge base and training necessary for prospective members of the profession. Accreditation is a guarantee to prospective students and employers that the academic program offers courses and experiences that graduates will need to succeed in the profession. All major professions have some form of accreditation of academic programs preparing young people to enter into the profession—medicine, law, engineering, and so on. Forestry also has an accreditation system. The Society of American Foresters (SAF), as the only professional society for the forestry community in the United States, has conducted periodic accreditation reviews of two- and four-year undergraduate forestry programs in the United States since 1935. Accreditation lasts for ten years.

An accreditation team visits the campus as part of the review to interview faculty, staff, and students. The team also examines course syllabi and facilities. Faculty members of the academic program are required to conduct a self-assessment as part of this review. The accreditation team reviews eight standard areas:

- Forestry program mission, goals, and objectives
- Curriculum
- Organization and administration of the forestry program
- Faculty
- Students
- Parent institution support
- Physical resources and facilities
- Research, extension, continuing education, and public service

The schools listed here have been accredited by SAF for the past ten years. The list includes schools that offer both two-year and four-year academic programs.

Auburn University
108 M. White Smith Hall
Auburn University
Auburn, AL 36849-5418
Phone: (334) 844-1007
E-mail: brinker@
forestry.auburn.edu
Web site: http://www.forestry.
auburn.edu

California Polytechnic State University
1 Grand Avenue
San Luis Obispo, CA 93407
Phone: (805) 756-2702
E-mail: dpiirto@calpoly.edu
Web site:
http://nrm.calpoly.edu

Clemson University
132 Lehotsky Hall
P.O. Box 340306
Clemson, SC 29634-0306
Phone: (864) 656-3215
E-mail: adunn@clemson.edu
Web site: http://www.
  clemson.edu/natres

Colorado State University
101 Natural Resource Building
Fort Collins, CO 80523
Phone: (970) 491-6911
E-mail: redente@
  cnr.colostate.edu
Web site: http://www.cnr.
  colostate.edu

Duke University
P.O. Box 90329
Durham, NC 27708-0329
Phone: (919) 613-8004
E-mail: drichter@duke.edu
Web site: http://www.env.
  duke.edu/communicator.
  html

Humboldt State University
101 Forestry Building
Arcata, CA 95521
Phone: (707) 826-3256
E-mail: cnrsdean@laurel.
  humboldt.edu
Web site: http://www.
  humboldt.edu/~cnrs

Iowa State University
251 Bessey Hall
Ames, IA 50011
Phone: (515) 294-1166

E-mail: jmkelly@iastate.edu
Web site: http://www.ag.
  iastate.edu/departments/
  forestry/Forestry.html

Louisiana State University
227 Forestry-Wildlife-Fisheries
  Building
Baton Rouge, LA 70803-6200
Phone: (225) 578-4167
E-mail: jchamb@lsu.edu
Web site: http://www.
  fwf.lsu.edu

Michigan State University
126 Natural Resource Building
East Lansing, MI 48824-1222
Phone: (517) 355-0093
E-mail: keathley@
  pilot.msu.edu
Web site: http://www.
  for.msu.edu

Michigan Tech University
1400 Townsend Drive
Houghton, MI 49931
Phone: (906) 487-2454
E-mail: gdmroz@mtu.edu
Web site: http://forestry.
  mtu.edu

Mississippi State University
109 Thompson Hall
P.O. Box 9680
Mississippi State, MS 39762-
  9690
Phone: (662) 325-8726
E-mail: bkarr@cfr.msstate.edu
Web site: http://www.
  cfr.msstate.edu

North Carolina State University
P.O. Box 8008
3019 Jordan Hall
Raleigh, NC 27695-8008
Phone: (919) 515-7789
E-mail: fred_cubbage@ncsu.edu
Web site: http://www.cfr.ncsu.edu/for/

Northern Arizona University
Box 15018
Flagstaff, AZ 86011-5018
Phone: (928) 523-6636
E-mail: bruce.fox@nau.edu
Web site: http://www.for.nau.edu/forestry

Ohio State University
2021 Coffey Road
Columbus, OH 43210
Phone: (614) 292-2265
E-mail: mullins.2@osu.edu
Web site: http://www.ag.ohio-state.edu/~natres

Oklahoma State University
008C Agriculture Hall
Stillwater, OK 74078-6013
Phone: (405) 744-5438
E-mail: mckincr@okstate.edu
Web site: http://www.okstate.edu/OSU_Ag/asnr/fore

Oregon State University
Peavy Hall 150
Corvallis, OR 97331-5704
Phone: (541) 737-1585
E-mail: hal.salwasser@orst.edu

Web site: http://www.cof.orst.edu

Pennsylvania State University
113 Ferguson Building
University Park, PA 16802-4300
Phone: (814) 863-7093
E-mail:chs3@psu.edu
Web site: http://www.cas.psu.edu/docs/CASDEPT/FOREST/homepage.htm

Purdue University
1159 Forestry Building
West Lafayette, IN 47907-1159
Phone: (765) 494-3590
E-mail: dclmstr@fnr.purdue.edu
Web site: http://www.fnr.purdue.edu

Southern Illinois University
Mailcode 4411
Carbondale, IL 62901-4411
Phone: (618) 453-3341 or (618) 453-7464
E-mail: jphelps@siu.edu
Web site: http://www.siu.edu/~forestry/forest.html

Stephen F. Austin State University
Box 6109, SFA Station
Nacogdoches, TX 75962-6109
Phone: (409) 468-3304
E-mail: sbeasley@sfasu.edu

Web site: http://www.
environment.sfasu.edu

**SUNY College of
Environmental Science and
Forestry**
320 Bray Hall
1 Forestry Drive
Syracuse, NY 13210
Phone: (315) 470-6534
E-mail: wbentley@esf.edu
Web site: http://www.esf.
edu/faculty/for/

Texas A&M University
Horticulture/Forest Science
Building
College Station, TX 77843-2135
Phone: (409) 845-5000
E-mail: tat-smith@tamu.edu
Web site: http://agrinet.
tamu.edu/FOREST/
department

**University of Alaska–
Fairbanks**
P.O. Box 757140
Fairbanks, AK 99775-7140
Phone: (907) 474-7083
E-mail: ffcel@uaf.edu
Web site: http://www.uaf.
edu/catalog/current/
schools/agriculture.html

**University of Arkansas–
Monticello**
P.O. Box 3468
Monticello, AR 71655
Phone: (870) 460-1052
E-mail: kluender@uamont.edu
Web site: http://www.afrc.
uamont.edu/sfr/index.htm

**University of California–
Berkeley**
145 Mulford Hall
Berkeley, CA 94720-3114
Phone: (510) 643-5428
E-mail: standifo@nature.berke-
ley.edu
Web site: http://www.cnr.
berkeley.edu/
departments/espm/

University of Florida
118 Newins-Ziegler Hall
P.O. Box 110410
Gainseville, FL 32611-0410
Phone: (352) 846-0850
E-mail: whs@gnv.ifas.ufl.edu
Web site: http://www.sfrc.
ufl.edu

University of Georgia
Forest Resources Building,
Room 2-229
Athens, GA 30602-2152
Phone: (706) 542-7247
E-mail: bongarte@
smokey.forestry.uga.edu
Web site: http://www.uga.
edu/~wsfr

University of Idaho
College of Natural Resources
Moscow, ID 83844-1133
Phone: (208) 885-7311
E-mail: joellen@uidaho.edu
Web site: http://www.uidaho.
edu/cfwr/fr/fr.html

University of Illinois
West 503 Turner Hall
Urbana, IL 61801
Phone: (217) 333-2770

E-mail: g-rolfe@uiuc.edu
Web site: http://www.aces.
uiuc.edu/~nres

University of Kentucky
205 Thomas Poe Cooper
Building
Lexington, KY 40546-0073
Phone: (606) 247-7496
E-mail: dgraves@ca.uky.edu
Web site: http://www.uky.
edu/Agriculture/Forestry/
forestry.html

University of Maine
5755 Nutting Hall
Orono, ME 04469-5755
Phone: (207) 581-2856
E-mail:
field@umenfa.maine.edu
Web site: http://www.ume.
maine.edu/~nfa/for_mgt/
welcome.htm

University of Massachusetts
Holdsworth Natural
Resources Center
Box 34210
Amherst, MA 01003-4210
Phone: (413) 545-1764
E-mail: bmccomb@
forwild.umass.edu
Web site: http://www.umass.
edu/forwild

University of Minnesota
115 Green Hall
1530 Cleveland Avenue N.
St. Paul, MN 55108
Phone: (612) 624-3400
E-mail: aek@forestry.umn.edu

Web site: http://www.cnr.
umn.edu/FR

University of Missouri
203 Anheuser-Busch Natural
Resources Building
Columbia, MO 65211
Phone: (573) 882-2627
E-mail: settergrenc@
missouri.edu
Web site: http://www.snr.
missouri.edu

University of Montana
109 Forestry Building
Missoula, MT 59812
Phone: (406) 243-5522
E-mail: pbrown@
forestry.umt.edu
Web site: http://www.forestry.
umt.edu

**University of New
Hampshire**
215 James Hall
Durham, NH 03824-3589
Phone: (603) 862-1020
E-mail: bill.mcdowell@
unh.edu
Web site: http://www.unh.
edu/natural-resources/
index.html

University of Tennessee
274 Ellington Plant Science
Building
Knoxville, TN 37996-4563
Phone: (865) 974-7126
E-mail: ghopper@utk.edu
Web site: http://fwf.ag.
utk.edu/

University of Vermont
Aiken Center
Burlington, VT 05405
Phone: (802) 656-4280
E-mail: ddehayes@
nature.snr.uvm.edu
Web site: http://nature.
snr.uvm.edu

University of Washington
Box 352100
Seattle, WA 98195
Phone: (206) 685-1928
E-mail: bare@
u.washington.edu
Web site: http://www.
cfr.washington.edu/

University of Wisconsin
1630 Linden Drive
Madison, WI 53706-3976
Phone: (608) 262-3976
E-mail: jcstier@
facstaff.wisc.edu
Web site: http://forest.
wisc.edu

**University of Wisconsin–
Madison**
146 Agricultural Hall
1450 Linden Drive
Madison, WI 53706
Phone: (608) 262-6968
E-mail: kmcsween@
facstaff.wisc.edu
Web site: http://www.
cals.wisc.edu/snr/

**University of Wisconsin–
Stevens Point**
1900 Franklin Street
Stevens Point, WI 54481
Phone: (715) 346-4617
E-mail: vphillip@uwsp.edu
Web site: http://www.
uwsp.edu/cnr

Utah State University
5200 Old Main Hill
Logan, UT 84322-5215
Phone: (435) 797-3219
E-mail: tlsharik@cc.usu.edu
Web site: http://www.
cnr.usu.edu/

Virginia Tech
307 Cheatham Hall
Blacksburg, VA 24061-0324
Phone: (540) 231-5483
E-mail: burkhart@vt.edu
Web site: http://www.
fw.vt.edu/forestry

West Virginia University
P.O. Box 6125
Morgantown, WV 26506
Phone: (304) 293-2941
Web site: http://www.caf.
wvu.edu/for/index.html

Yale University
205 Prospect Street
New Haven, CT 06511
Phone: (203) 432-5109
E-mail: gus.speth@yale.edu
Web site: http://www.
yale.edu/forestry/

Forest Products Companies

Boise Cascade
1111 West Jefferson Street
P.O. Box 50
Boise, ID 83728-0001
Phone: (208) 384-6161
E-mail: bcweb@bc.com
Web site: http://www.bc.com

Boise is an international distributor of office supplies and paper and an integrated manufacturer and distributor of paper, packaging, and building materials. Its operations extend throughout North America, Australia, New Zealand, and Brazil. It owns or manages 2.3 million acres of timberlands in the United States to support its operations.

Bowater Inc.
55 East Camperdown Way
Greenville, SC 29602
Phone: (864) 271-7733
Web site: www.bowater.com/

Bowater Inc., headquartered in Greenville, South Carolina, makes newsprint, coated and uncoated groundwood papers, bleached kraft pulp, and lumber products, and it converts a groundwood base sheet to coated products.

Collins Companies
1618 S.W. First Avenue
Suite 500
Portland, OR 97201
(800) 329-1219
(503) 227-1219
Web site: http://www.collinswood.com

The family-owned Collins Companies traces its roots back to 1855 in Pennsylvania and has expanded west to manage 94,000 acres of softwood in northeastern California. Today its holdings also include the Collins Lakeview Forest in southern Oregon and northern California. Collins Products LLC manufactures standard and certified exterior siding and particle board and standard hardboard siding and trim.

Duluth Timber Company
P.O. Box 16717
Duluth, MN 55816
Web site: http://www.duluthtimber.com

Duluth Timber Company, using reclaimed or salvaged wood, is dedicated to satisfying the needs of new projects with old wood.

Georgia Pacific
133 Peachtree Street NE
Atlanta, GA 30303
Phone: (404) 652-4000
Web site: http://www.gp.com

Georgia Pacific manufactures and distributes tissue, packaging, paper, building products, and pulp and related chemicals. Headquartered in Atlanta, Georgia-Pacific employs more than 75,000 people at 600 locations worldwide.

Home Depot
Consumer Affairs
2455 Paces Ferry Road
Atlanta, GA 30339
Phone: (770) 433-8211
Web site: http://www.homedepot.com

Home Depot is the largest seller of lumber in the United States.

Mead Westvaco Corporation
1 High Ridge Park
Stamford, CT 06905
Phone: (203) 461-7400
Web site: http://www.mead.com

MeadWestvaco Corporation has annual sales of $8 billion and produces packaging, coated and specialty papers, consumer and office products, and specialty chemicals. MeadWestvaco operates in 33 countries, serves customers in approximately 100 nations, employs more than 30,000 people worldwide, and owns 3.4 million acres of forests managed using sustainable forestry practices.

Universal Forest Products Inc.
2801 East Beltline NE
Grand Rapids, MI 49525
Phone: (616) 364-6161
Web site: http://www.ufpi.com

Universal Forest Products Inc. is the nation's leading manufacturer and distributor of wood products for retail home centers, residential and commercial site-built construction, the industrial market, and structural lumber products for the manufactured housing industry.

WNC Pallet and Forest Products Co.
P.O. Box 38
1414 Smoky Park Highway
Candler, NC 28715
Phone: (828) 667-5426
E-mail: wncpallet@bellsouth.net
Web site: http://www.wncpallet.com

WNC Pallet and Forest Products is a forest products supplier in western North Carolina. It buys land and timber and operates a modern sawmill complete with planing, resawing, and dry-kiln capabilities. In addition, it produces pallets, skids, and boxes.

Forestry Trade Associations

American Forest and Paper Association
1111 Nineteenth Street NW
Suite 800
Washington, DC 20036
Phone: (202) 463-2700
E-mail: info@afandpa.org
Web site: http://www.afandpa.org

AF&PA is the national trade association of the forest, paper, and wood products industry. It represents member companies engaged in growing, harvesting, and processing wood and wood fiber; manufacturing pulp, paper, and paperboard products from both virgin and recycled fiber; and producing engineered and traditional wood products. It is a nonprofit education and conservation organization working for healthy forests and quality environmental education.

Composite Panel Association
18928 Premier Court
Gaithersburg, MD 20879-1569
Phone: (301) 670-0604

The Composite Panel Association is the national trade association for the particle board and medium density fiberboard industries. It is dedicated to increasing the acceptance and use of industry products and to providing for the general welfare of the industry.

Forest Products Society
2801 Marshall Court
Madison, WI 53705-2295
Phone: (608) 231-1361
E-mail: info@forestprod.org
Web site: http://www.forestprod.org

The Forest Products Society is an international not-for-profit technical association that provides an information network for all segments of the forest products industry. Its mission is to foster innovation and research in the environmentally sound processing and use of wood and fiber resources by disseminating information and providing forums for networking and the exchange of knowledge.

Forest Resources Association Inc.
600 Jefferson Plaza
Suite 350
Rockville, MD 20852
Phone: (301) 838-9385
Web site: http://www.apulpa.org

The mission of the Forest Resources Association is to promote the best interests of wood fiber suppliers and consumers in the economical, efficient, and sustainable use and stewardship of forest resources to meet wood fiber needs through private enterprise.

Hardwood Plywood and Veneer Association
P.O. Box 2789
Reston, VA 20195-0789
Phone: (703) 435-2900
E-mail: hpva@hpva.org
Web site: http://www.hpva.org

The HPVA represents the interests of the hardwood plywood, hardwood veneer, and engineered hardwood flooring industries. HPVA member companies produce nearly all of the hardwood plywood stock panels and sliced hardwood veneer manufactured in North America. It offers information and resources on hardwood, plywood veneer, and engineered flooring.

International Wood Products Association
4214 King Street
West Alexandria, VA 22302
Phone: (703) 820-6696
E-mail: info@iwpawood.org
Web site: http://www.iwpawood.org

IWPA's mission is to advance international trade in wood products through leadership in business, environmental, and government affairs. The association serves as the North American imported wood products industry's "spokesgroup" at all levels of government.

National Hardwood Lumber Association
6830 Raleigh-LaGrange Road
Memphis, TN 38184-0518
Phone: (901) 377-1818
E-mail: info@ntlhardwood.org
Web site: http://www.ntlhardwood.org

The mission of NHLA is to maintain order, structure, and ethics in the hardwood marketplace; provide specialized member services unique to the hardwood industry; assure timber availability to meet society's needs; and build positive relationships within the hardwood community.

Technology Association for the Pulp and Paper Industries
P.O. Box 105113
Atlanta, GA 30348

Phone: (770) 446-1400
Web site: http://www.tappi.org

The Technology Association for the Pulp and Paper Industries (TAPPI) is the leading technical association for the worldwide pulp and paper industry. TAPPI provides its members quick access to the largest international group of technically experienced people in the industry, the most comprehensive collection of reliable technical information and knowledge in the industry, and the highest-quality products and services created to meet the needs of people who solve industry technical problems.

Forestry Professional Associations

American Fisheries Society
5410 Grosvenor Lane
Bethesda, MD 20814
Phone: (301) 897-8616
E-mail: main@fisheries.org
Web site: http://www.fisheries.org

The mission of the American Fisheries Society is to improve the conservation and sustainability of fishery resources and aquatic ecosystems by advancing fisheries and aquatic science and promoting the development of fisheries professionals. Much fish habitat relies on stream shading provided by trees.

Association of Consulting Foresters of America Inc.
732 N. Washington Street
Suite 4-A
Alexandria, VA 22314-1921
Phone: (703) 548-0990
E-mail: director@acf-foresters.com
Web site: http://www.acf-foresters.com

The ACF was founded to advance the professionalism, ethics, and interests of professional foresters who primarily consult to the public. ACF is active in setting the standards for the consulting forestry profession; educating and assisting landowners in good forest stewardship; tracking legislation that directly affects private forestry; educating the public, legislators, and others about issues sensitive to private landowners; and cooperating with Federal and state agencies to protect private property rights.

Forest History Society
701 William Vickers Avenue
Durham, NC 27701-3162
Phone: (919) 682-9319
Web site: http://www.lib.duke.edu/forest/

The Forest History Society links the past to the future by identifying, collecting, preserving, interpreting, and disseminating information on the history of interactions between people, forests, and their related resources—timber, water, soil, forage, fish and wildlife, recreation, and scenic or spiritual values.

Society of American Foresters
5400 Grosvenor Lane
Bethesda, MD 20814
Phone: (301) 897-8720
E-mail: safweb@safnet.org
Web site: http://www.safnet.org

The Society of American Foresters (SAF) is the national scientific and educational organization representing the forestry profession in the United States. The mission of the SAF is to advance the science, education, technology, and practice of forestry; to enhance the competency of its members; to establish professional excellence; and to use the knowledge, skills, and conservation ethic of the profession to ensure the continued health and use of forest ecosystems and the present and future availability of forest resources to benefit society.

Wildlife Society
5410 Grosvenor Lane
Suite 200
Bethesda, MD 20814-2144
Phone: (301) 897-9770
Web site: http://www.wildlife.org

The Wildlife Society is an international nonprofit scientific and educational association dedicated to excellence in wildlife stewardship through science and education. Its mission is to enhance the ability of wildlife professionals to conserve diversity, sustain productivity, and ensure responsible use of wildlife resources for the benefit of society.

Other Forestry-Related Associations

American Forest Foundation
1111 Nineteenth Street NW
Suite 780
Washington, DC 20036
Phone: (202) 463-2462
E-mail: info@forestfoundation.org
Web site: http://www.affoundation.org

The American Forest Foundation strives to promote the growing of renewable forest resources on private lands while protecting environmental benefits and increasing public understanding of all benefits of productive forestry.

American Forest Resource Council
1500 S.W. First Avenue
Suite 330
Portland, OR 97201
Phone: (503) 222-9505
Web site: http://www.afrc.ws

AFRC strives to create a favorable operating environment for the forest products industry; ensure a reliable timber supply from public and private lands; and promote sustainable management of forests by improving Federal laws, regulations, policies, and decisions that determine or influence the management of all lands.

American Tree Farm System
c/o American Forest Foundation
1111 Nineteenth Street NW
Suite 780
Washington, DC 20036
Phone: (202) 463-2462
E-mail: info@treefarmsystem.org
Web site: http://www.treefarmsystem.org

The American Tree Farm System® (ATFS), a program of the American Forest Foundation, is committed to sustaining forests, watersheds, and healthy habitats through the power of private stewardship. ATFS educates and recognizes the commitment of private forest owners in the United States.

Certified Forest Products Council
721 NW Ninth Avenue
Suite 300
Portland, OR 97209
Phone: (503) 224-2205
E-mail: info@certifiedwood.org
Web site: http://www.certifiedwood.org

The Certified Forest Products Council is an independent, not-for-profit, voluntary initiative committed to promoting responsible forest products–buying practices throughout North America in an effort to improve forest management practices worldwide. The council actively promotes and facilitates the increased purchase, use, and sale of certified forest products and promotes the transition away from forest products originating in forests that have been identified as endangered through an assessment process.

Forest Stewards Guild
P.O. Box 8309
Santa Fe, NM 87504-8309
Phone: (505) 983-3887
E-mail: info@foreststewardsguild.com
Web site: http://foreststewardsguild.com

The mission of the guild is to promote ecologically responsible resource management that sustains the entire forest across the landscape. The guild provides a forum and support system for practicing foresters and other resource management professionals working to advance this vision.

Forest Stewardship Council, U.S.
1155 Thirtieth Street NW
Suite 300
Washington, DC 20007
Phone: (202) 342-6589
Web site: http://fscus.org/html/index.html

The FSC promotes responsible forest management globally by certifying forest products that meet the most rigorous standards in the world. The organization brings industry, environmentalists, and community groups together to promote practical solutions that meet its diverse stakeholders' needs.

Global Association of Online Foresters
Web site: http://www.foresters.org/

The aim of the Global Association of Online Foresters (GAOF) is to increase the understanding of forestry by improving communication worldwide. Services include information links, a help line, assistance with careers including job searching, and a registry of consultants. The GAOF also helps members promote forestry in their respective countries by providing directories of e-mail addresses and web sites.

National Association of Professional Forestry Schools and Colleges
701 Pennsylvania Avenue NW
Washington, DC 20004
Phone: (202) 624-1280
Web site: http://www.napfsc.org

The association was formed to advance the science, practice, and art of forest resource management through the encouragement and support of forest resource education, research, extension, and international programs at the university level. NAPFSC represents member schools and institutions in their collective dealings with governmental entities, forestry and other natural resource organizations, forestry industry groups, educational and scientific organizations, and the public.

National Network of Forest Practitioners
305 Main Street
Providence, RI 02903
Phone: (401) 273-6507
E-mail: Thomas@nnfp.org
Web site: http://www.nnfp.org

The National Network of Forest Practitioners is a grassroots alliance of rural people who are striving to build a forest economy that is ecologically sound and socially just. As one of the leading community forestry organizations in the United States, the NNFP provides information and technical assistance, a forum for networking and organizing, and a meaningful role in national discussions about forests and rural communities.

National Tree Trust
1120 G Street NW
Suite 770
Washington, DC 20005
Phone: (202) 628-8733
Web site: http://www.nationaltreetrust

The mission of the National Tree Trust is to act as a catalyst for local volunteer groups in the growing, planting, and maintenance of trees in rural communities, urban areas, and along the nation's highways.

National Woodland Owners Association
374 Maple E
Suite 310
Vienna, VA 22180
Phone: (703) 255-2700
E-mail: info@woodlandowners.org
Web site: http://www.woodlandowners.org

The National Woodland Owners Association (NWOA) is a nationwide organization made up of nonindustrial private woodland owners. NWOA is independent of the forest products industry and government forestry agencies. NWOA works with all organizations to promote nonindustrial forestry and the best interests of woodland owners. Its purpose is to develop policy, legislation, and representation at the national level as one unified voice as well as to provide educational and networking opportunities to landowners throughout the country.

7

Print Resources

There is an enormous amount of printed information about forestry and forest conservation policy in the United States. Writings about forest conservation policy date back to the late 1800s, with key figures such as Carl Schenck, Bernhard Fernow, and Gifford Pinchot calling for a more coherent policy approach to forest conservation to counter the "cut-and-run" practices of the time. Forest conservation policy again became a focal point in the 1950s with the emergence of multiple-use forestry. But it wasn't until the mid-1970s, with the passage of the National Forest Management Act of 1976 and a spate of state-level forest practices laws, when forest conservation policy became a field unto its own in the forestry world. Several important books were written about the challenges inherent in balancing the multiple demands on forests, from timber products to water quality and recreation. Public involvement also became a pervasive theme.

The conflicts over the northern spotted owl in the Pacific Northwest during the late 1980s and early 1990s brought forest conservation policy within the broader fold of biodiversity. The biodiversity issue was brought into sharper focus with news reports about tropical deforestation in Brazil and other parts of the world. By 2000, forest conservation policy had become synonymous with biodiversity conservation in the literature.

Much of the focus in forest policy books and forestry journals has been on public forest land, especially targeting the U.S. Forest Service. As the steward of 192 million acres of public forest land, the Forest Service is under constant pressure to provide for human uses as well as conserving biodiversity and other ecosystem goods and services. Far less attention has been given to private forest conservation, even though private forests comprise 58 percent of the total forest land in the United States.

Forest science and research journals are also abundant, a testament to the importance of forests to society at large. Scientific journals are generally inclusive, with topics ranging from forest genetics and soil microbiology to economic models and policy analysis.

General Forestry and Forest Conservation Books and Reports

Aplet, G. H., N. Johnson, J. T. Olsen, and V. A. Sample, eds. 1993. *Defining Sustainable Forestry.* Covelo, CA: Island Press. 328p.

A broad collection of essays on the meaning of sustainable forestry by authors from varying perspectives. Contributors range from conservation biologists, academic researchers, and Forest Service decision-makers to forest industry representatives, environmentalists, and nonindustrial private landowners. The essays reflect the rich and diverse perspectives on what constitutes sustainable forestry.

Best, C., and L. A. Wayburn. 2001. *America's Private Forests: Status and Stewardship.* Covelo, CA: Island Press.

A comprehensive overview of status and trends in U.S. private forest land ownership and uses. The authors concentrate on constructive recommendations to enhance the conservation, restoration, and sustainable management of these lands.

Birch, T. W. 1994. *Private Forest-Land Owners of the United States, 1994.* Resource Bulletin NE-134. Radnor, PA: U.S. Department of Agriculture, U.S. Forest Service, Northeastern Forest Experiment Station. 183p.

An indispensable source of data on changes in private forest land ownership between 1978 and 1994, including information about private land owner objectives for ownership. Data on ownership acreage are organized by ownership category, as well as by region and state.

Davis, L. S., and K. N. Johnson. 1986. *Forest Management, 3rd Edition.* New York: McGraw-Hill.

The fundamental textbook on forest management used by accredited forestry schools in the United States for many years. The focus of the text is on maximizing the production of forest values, focusing primarily on timber but also including other examples.

Forest Ecosystem Management Assessment Team (FEMAT). 1993. *Forest Ecosystem Management: An Ecological, Economic, and Social Assessment.* Report of FEMAT, 1993-793-071. Washington, DC: General Printing Office.

A comprehensive report by a team of forest ecologists, economists, and social scientists assembled in 1993 by the Clinton Administration to provide analysis and recommendations to manage a part of the Federal forest lands in the Pacific Northwest and northwest California. The report was the first of its kind to analyze the effects of timber harvesting on Federal forest lands on a broad array of ecological and social values supported by old-growth forests in the region. The resulting analysis provided the foundation for the Northwest Forest Plan, which governs forest conservation and management in a large part of National Forests in the Pacific Northwest and northwest California.

Hunter, M. L., Jr., ed. 1999. *Maintaining Biodiversity in Forest Ecosystems.* Cambridge, England: Cambridge University Press.

This text is aimed at a broad audience concerned about the effects of forest management on biological diversity and how to adapt forestry to meet broader goals of biodiversity. The book includes basic concepts of biodiversity and forest ecology and provides constructive recommendations for a restoration-based forestry.

Kimmins, H. 1992. *Balancing Act: Environmental Issues in Forestry.* Vancouver: University of British Columbia Press.

Basic principles of forestry and ecology are outlined in the beginning of the book, followed by major issues facing forestry in British Columbia and worldwide. The issues include clearcutting, slash-burning, chemicals, old-growth forests, biological diversity, "new forestry" initiatives, climate change, acid rain, decision-making, tropical forestry, and sustainability of diverse forest values.

Knight, R. L., and S. F. Bates, eds. 1995. *A New Century for Natural Resources Management.* Covelo, CA: Island Press. 398p.

A collection of writings by leaders in U.S. natural resource conservation and management that reflect on the dysfunctions, challenges, and opportunities for the twenty-first century. A primary thrust of the writings is the philosophies and principles underlying sustainable natural resource management.

Kohm, K. A., and J. F. Franklin, eds. 1997. *Creating a Forestry for the 21st Century: The Science of Ecosystem Management.* Covelo, CA: Island Press. 475p.

A series of highly technical chapters concerning the integration of ecological principles into forest management and conservation. The chapters are organized into three general themes: principles of ecosystem management across spatial and temporal scales; applications of forest ecosystem management; and social, institutional, and economic issues surrounding the implementation of— and impediments to—forest ecosystem management.

Ricketts, T. H., E. Dinerstein, D. M. Olson, and C. J. Loucks. 2000. *Terrestrial Ecoregions of North America: A Conservation Assessment.* Washington, DC: Island Press.

An impressive collection of color maps and narratives of the status and trends of the ecoregions of North America.

Szaro, R. C., and B. Shapiro. 1990. *Conserving Our Heritage: America's Biodiversity.* Arlington, VA: The Nature Conservancy.

An assessment of biodiversity in the United States and recommendations for protecting, maintaining, and enhancing biodiversity into the future.

Yaffee, S. L., A. F. Phillips, I. C. Frentz, P. W. Hardy, S. M. Maleki, and B. E. Thorpe. 1996. *Ecosystem Management in the United States.* Covelo, CA: Island Press.

A comprehensive reference guide to ecosystem management efforts in the United States targeted at practitioners and decision-makers. The guide provides findings from 105 ecosystem management projects nationwide.

U.S. Forest Policy Books and Reports

Adams, D. 1993. *Renewable Resource Policy.* Washington, DC: Island Press.

A thorough survey of policies, programs, and key organizations affecting the use, management, and conservation of renewable natural resources including forests, rangeland, water, and wildlife.

Anderson, T. L. 1994. *Multiple Conflicts over Multiple Uses.* Bozeman, MT: Political Economy Research Center. 103p.

A critical analysis of how the government addresses conflicts over public lands and natural resources. Recommendations based on free-market principles are provided, with calls for reforms of the political and institutional environment surrounding public land and resource management.

Behan, R. W. 2001. *Plundered Promise: Capitalism, Politics, and the Fate of Federal Lands.* Covelo, CA, and Washington, DC: Island Press. 240p.

A critical analysis of the relationship between corporations and the U.S. Forest Service and the resulting timber program in National Forests. The central thesis is that current institutions have overshot their original mandates and missions, and that these mandates and missions need to be readdressed through an alternative model of forest sustainability, one that emphasizes community-based forest conservation and management.

Bowes, M., and J. Krutilla. 1989. *Multiple-Use Management: The Economics of Public Forestlands.* Washington, DC: Resources for the Future.

Basic concepts and principles of public economics applied to the allocation and management of public forest lands.

Clarke, J. N., and D. C. McCool. 1996. *Staking the Terrain: Power and Performance among Natural Resource Agencies, 2nd Edition.* Albany, NY: State University of New York Press. 279p.

An analysis of Federal natural resource agencies and their relative power and effectiveness. The Forest Service is regarded as a bureaucratic superstar. The National Park Service, Bureau of Land Management, Fish and Wildlife Service, and Army Corps of Engineers are also examined.

Clary, D. A. 1986. *Timber and the Forest Service.* Lawrence: University of Kansas Press. 252p.

An interpretive history of the relationship between timber management and Forest Service organization, planning, and decision-making. Historical documents, official memos, reports, and internal analyses are placed under a critical light to show how the Forest Service evolved from being a custodial steward of Federal forest land into a timber production organization in a very short period of time. How the Forest Service extracts itself from predominantly timber production to multiple use and values remains a persistent challenge.

Clawson, M. 1975. *Forests for Whom and for What?* Baltimore, MD: Johns Hopkins University Press for Resources for the Future.

Data and trends of National Forest use and management are examined against political demands on National Forests. Clawson raises critical questions about what the future might hold and what possible directions the Forest Service might take to manage the National Forests into the future.

Clawson, M. 1983. *The Federal Lands Revisited.* Baltimore, MD: Johns Hopkins University Press for Resources For the Future.

Data and trends for all Federal lands, National Forests, Bureau of Land Management lands, national parks, and national wildlife refuges.

Committee of Scientists. 1999. *Sustaining the People's Lands: Recommendations for Stewardship of the National Forests and Grasslands into the Next Century.* Washington, DC: U.S. Department of Agriculture, U.S. Forest Service.

The Committee of Scientists was appointed by Congress in 1998 to provide analysis and recommendations to the Forest Service in revising the administrative regulations for planning and management of the National Forests and grasslands. This report emphasizes ecological sustainability over social and economic sustainability by taking an ecosystem management approach to National Forest planning and management. The report also calls for more collaborative approaches and sustained public involvement throughout decision-making processes.

Cortner, H. J., and M. A. Moote. 1999. *The Politics of Ecosystem Management.* Covelo, CA: Island Press. 179p.

This book examines ecosystem management in the context of political theory and principles, from the "politics of expertise" and policy paradoxes of ecosystem management to new forms of public participation in land and natural resource decision-making. The authors critically analyze the new role of scientists, managers, and the public in putting ecosystem management principles into practice.

Cubbage, F. W., J. O'Laughlin, and C. S Bullock. 1993. *Forest Resource Policy.* New York: Wiley.

Intended as a textbook for introductory courses in forest policy, *Forest Resource Policy* covers a broad range of policies and participants affecting forest conservation and management in the United States. Early chapters focus on the various values provided by and assigned to forests and the role of policy in ensuring that those values are sustained. Middle chapters focus on key participants in forest policy and management, such as the legislative, executive, and judicial branches; interest groups; the public; and the media. Major laws and policies affecting both public and private forest lands follow. The text ends with discussions of international forestry and forest policy.

Dana, S., and S. K. Fairfax. 1980. *Forest and Range Policy: Its Development in the United States, 2nd Edition.* New York: McGraw-Hill.

The original textbook intended for introductory courses in forest and range policy. The text takes a historical narrative approach, tracking the development and implementation of policies through major historical periods and events up to 1980. The text doubles as a good historical primer on forest conservation policy.

Gray, G. J., M. J. Enzer, and J. Kusel. 1998. *Understanding Community-Based Forest Ecosystem Management.* An editorial synthesis of an American Forests workshop, Bend, Oregon, June 1998. Washington, DC: American Forests.

A synthesis of papers given at a national conference on community-based forest ecosystem management. Basic principles and essential practices are described and applied in the context of sustainable forestry and sustainable communities.

Hirt, P. 1996. *A Conspiracy of Optimism: Management of the National Forests Since World War Two.* Lincoln: University of Nebraska Press.

This book tracks the emergence of timber production as an embodiment of technical efficiency and optimistic attitudes that pervaded the Forest Service after World War II. The author criticizes the agency as a "can-do" organization that refused to recognize the long-term negative effects of the timber program, despite growing scientific evidence and public opposition to the program.

O'Toole, R. 1988. *Reforming the Forest Service.* Covelo, CA: Island Press. 247p.

An analysis of Forest Service planning and decision-making processes and priorities in the context of public economic theory. The Forest Service acts as any "rent-seeking" actor by constantly looking for opportunities to maximize its budgets regardless of the negative environmental, economic, and social effects. O'Toole suggests restructuring the Forest Service to operate under free market principles and practices, which emphasize efficiency and equity to a greater extent than the current model.

Sample, V. A. 1990. *The Impact of the Federal Budget Process on National Forest Planning.* Westport: Greenwood Press.

This is a compilation of data and trends that couple Forest Service planning, decision-making, and management with annual congressional appropriations. Regardless of land allocations, conservation strategies, and management priorities outlined in forest plans, the agency implements actions in correspondence to its budget structure. Line items and expanded budget line items track closely with on-the-ground management much more closely than forest plans. Reform of the Forest Service begins and ends with the federal budget process.

U.S. Department of Agriculture (USDA), U.S. Forest Service. 1993. *The Principal Laws Relating to Forest Service Activities.* Washington, DC: U.S. Government Printing Office.

A collection of all congressional statutes relating to Forest Service activities.

Wilkinson, C. F., and H. M. Anderson. 1987. *Land and Resource Planning on the National Forests.* Covelo, CA: Island Press.

History, detailed description, and execution of the Renewable Resources Planning Act of 1974 and the National Forest Management Act of 1976.

Wondolleck, J. M. 1988. *Public Lands Conflict and Resolution: Managing National Forest Disputes.* New York: Plenum Press.

In-depth examination of public involvement and dispute-resolution approaches in National Forest planning and management decision-making. The examination applies principles of effective public involvement and dispute resolution against empirical information from National Forest disputes. In conclusion, the Forest Service can be more effective if it engages public stakeholders earlier in the decision process and more often throughout the process.

Wondolleck, J. M., and S. J. Yaffee. 2000. *Making Collaboration Work: Lessons from Innovation in Natural Resource Management.* Covelo, CA: Island Press.

This book presents a series of case studies and lessons learned from collaborative approaches to natural resource management in the United States. Lessons include maintaining a place-based focus, developing open and fair procedures, emphasizing social learning and conflict management, and instituting monitoring and evaluation to ensure adaptive management.

Yaffee, S. J. 1994. *The Wisdom of the Spotted Owl: Policy Lessons for a New Century.* Covelo, CA: Island Press.

Interpretive history and analysis of the policy conflicts over the protection of the northern spotted owl and old-growth forests on Federal lands in the Pacific Northwest. The first half of the book provides historical development of the issue; the second half places the issue in a broader political context.

U.S. Forest History

Clawson, M. 1967. *The Federal Lands since 1956: Recent Trends in Use and Management.* Baltimore, MD: Johns Hopkins University Press for Resources for the Future.

A comprehensive collection of data on the uses of all Federal lands through the mid-1960s. Data are for all management activities in national forests, Bureau of Land Management public lands, and national parks—for example, timber harvest volume, oil and gas leases, and recreation visitors.

Frederick, K. D., and R. A. Sedjo, eds. 1991. *America's Renewable Resources: Historical Trends and Current Challenges.* Washington, DC: Resources for the Future.

This book traces the history of the uses of water, forests, rangeland, cropland and soils, and wildlife since early colonial times through major periods and events marking the expansion and development of the United States. The authors analyze their current condition at the end of the twentieth century and the role of social and political institutions in perpetuating degradation or conservation.

MacCleery, D. W. 1996. *American Forests: A History of Resiliency and Recovery, 4th revision.* Durham, NC: Forest History Society.

This short handbook presents data and trends regarding U.S. forests, including information on total forest land acreage, forest growth and harvesting, and forest health conditions from colonial settlement through the late twentieth century.

Miller, C., and R. Staebler. 1999. *The Greatest Good: 100 Years of Forestry in America.* Bethesda, MD: Society of American Foresters.

A coffee table–type volume of interpretive photo essays on the history of U.S. forestry since 1900. Rare black-and-white photos from forestry's early beginnings and pictures representing today's forestry are collected with narratives and quotations from forestry leaders throughout the twentieth century.

Pinchot, G. 1907. *The Use of the National Forests.* Washington, DC: U.S. Department of Agriculture.

This handbook was distributed to the first generation of National Forest managers to guide their decisions regarding the uses of the newly created National Forests. The handbook contains the most famous of quotes attributed to Gifford Pinchot, first Chief of the Forest Service, to "provide for the greatest good, for the greatest number, over the long term."

Pinchot, G. 1911. *The Fight for Conservation.* New York: Doubleday, Page and Company.

Written by Gifford Pinchot, the first Chief of the Forest Service, this book contains the essence of the early Conservation Movement blended with essential principles of the Progressive Era reforms. Pinchot laid out a three-pronged approach to natural resource conservation: conservative use of resources for current generations, protection of resources from wasteful practices for future generations, and distribution and management of resources in a way that is just and equitable.

Pinchot, G. 1947. *Breaking New Ground.* New York: Harcourt, Brace and Company.

A collection of personal essays, journal excerpts, and philosophical reflections by the first chief of the Forest Service, Gifford Pinchot. The book is a highly personal account by a personality at the forefront of the Conservation Movement at the turn of the twentieth century and the social and political changes wrought by conservation policy and politics.

Pyne, S. J. 1982. *Fire in America: A Cultural History of Wildland and Rural Fire.* Seattle, WA: University of Washington Press.

An essential work on the history of wildland fire in the United States and social and political responses to fire. Detailed accounts of the famous forest fires of 1910 and the emergence of Smokey Bear enrich this narrative history of wildland fire.

Steen, H. K., ed. 1992. *The Origins of the National Forests.* Durham, NC: Forest History Society.

A comprehensive collection of essays on the emergence, management, and significance of the National Forests. Themes include the importance of National Forests; management of specific resources such as timber, mining, grazing, wildlife, and water; relationships between National Forests and Native Americans, states, and private lands; relationships between the Forest Service and other Federal agencies such as the National Park Service; and key personalities associated with the National Forests, such as Bernhard Fernow, John Waldo, and William Steel.

Williams, M. 1989. *Americans and Their Forests: A Historical Geography.* Cambridge, England: Cambridge University Press.

A dense analysis of the meaning and significance of forests in American history and culture, from the clearing and uses of forests in pre-European society to the late twentieth century. The analysis is compelling given the author's British perspective.

International Forestry and Forest Conservation

FAO (Food and Agriculture Organization). 2000. *Global Outlook for the Future Wood Supply from Forest Plantations.* Rome: United Nations Food and Agriculture Organization.

An analysis of trends in consumption of wood products and production of wood from forest plantations. As demand for wood products continuously grows, there is increasing interest in the development of intensively managed forest plantations to replace forest product harvesting and development from native forests around the world. Plantations are examined in light of the need to protect native forests for biodiversity and other ecological values.

Howard, S., and J. Stead. 2001. *The Forest Industry in the 21st Century.* London: World Wide Fund for Nature.

Information about past, current, and future trends of the forest products industry around the globe. Included in the assessment are analyses of trade, corporate structuring, and effects on conserving nontimber forest values such as biodiversity.

Lele, U., ed. 2002. *Managing a Global Resource: Challenges of Forest Conservation and Development.* New Brunswick, NJ: Transaction Publishers.

A general overview and selected case studies of challenges and opportunities in forest conservation around the world. Case examples come from Costa Rica, China, India, Cameroon, Indonesia, and Brazil.

Pagiola, S., J. Bishop, and N. Landell-Mills. 2002. *Selling Forest Environmental Services: Market-Based Mechanisms for Conservation and Development.* London: Earthscan Publications Ltd. 299p.

An unconventional analysis of using market mechanisms to conserve forests, such as selling watershed protection services, trading carbon credits to reduce greenhouse gases, and creating markets for biodiversity protection. Case examples from around the world are included.

Prasad, B., ed. 1999. *Biotechnology and Biodiversity in Agriculture/Forestry.* Plymouth, England: Science Publishers Inc.

A technical text about the technologies for genome and DNA analysis and cloning and their applications in agriculture, forestry, and protection of global biodiversity.

Ramakrishna, K., and G. M. Woodwell, eds. 1993. *World Forests for the Future: Their Use and Conservation.* New Haven, CT: Yale University Press.

A general overview of issues and challenges in sustaining forests and forest values around the world.

Sharma, N. P., ed. 1992. *Managing the World's Forests: Looking for Balance Between Conservation and Development.* Dubuque, IA: Kendall/Hunt Publishing.

A global perspective on approaches to balance forest conservation with development needs, interests, and concerns of both developed and developing countries, with an emphasis on poverty alleviation.

Westoby, J. 1989. *Introduction to World Forestry: People and Their Trees.* Oxford, England: Blackwell.

An introductory guide to the coevolution of trees, forests, and people around the world, emphasizing the significance of trees and forests to human development and well-being. The guide also provides overviews of forest conflicts around the world and assesses policies and programs aimed at conserving forests.

World Commission on Environment and Development. 1987. *Our Common Future.* New York: Oxford University Press.

Known informally as the "Brundtland Report" after Prime Minister Gro Harlem Brundtland of Norway, who chaired the commission, this book provides definitions and guidelines for sustainable development, balancing economic development with environmental sustainability. The book is regarded as the definitive source literature for sustainable development.

Scientific Journals

Advances in Horticulture and Forestry. Scientific Publishers.

An international journal focusing on the science and technology applied to horticulture and forestry.

American Forests. American Forests Association.

American Forests is the magazine about trees and forests for people who know and appreciate the many benefits of trees. Stories are written to entice a general audience to care about tree planting; content includes profiles of forest advocates, in-depth looks at current controversies, and stories about current research.

Arboricultural Journal: The International Journal of Urban Forestry. A B Academic Publishers.

This journal is a source for all aspects of urban forest science, management, and administration.

Australian Forestry. Institute of Foresters of Australia.

This journal is a comprehensive source for all aspects of forest science and forest management.

Biological Conservation. Elsevier Science.

Biological Conservation has as its main purpose the widest dissemination of original papers dealing with the preservation of wildlife and the conservation or wise use of biological and allied natural resources. It is concerned with plants and animals and their habitats in a changing and increasingly human-dominated biosphere—in fresh and salt waters as well as on land and in the atmosphere. *Biological Conservation* publishes field studies, analytical and modeling studies, and review articles.

Biotechnology in Agriculture and Forestry. Springer-Verlag, Journals.

The purpose of this international journal is to disseminate research and insights about the use, application, and effect of biotechnology on agricultural and forest production.

Boston College Environmental Affairs Law Review. Boston College Environmental Affairs Inc.

The *Boston College Environmental Affairs Law Review* is the nation's second-oldest law review dedicated solely to environmental law. Since its inception in 1971, *Environmental Affairs* has developed and maintained a national reputation as one of the country's leading environmental journals. Its staff consists of approximately 15 second- and 15 third-year students who publish four issues during the academic year. Each issue contains articles contributed by prominent outside authors as well as several student-written legal comments.

Canadian Forest Industries Journal. NRC Research Press.

Canadian Forest Industries Journal, published six times per year, is the journal of record for the Canadian Woodlands Forum and focuses exclusively on all aspects of logging from the stump to the mill gate.

Canadian Journal of Forest Research. National Research Council of Canada.

The journal features articles about silviculture, forest mensuration, harvesting, vegetation management, tree physiology, ecophysiology, dendrochronology, forest ecology, forest fire ecology, forest soil biology, biotechnology, forest genetics, tree improvement, forest entomology and pathology, effects of pollution and global change, effects of forest practices on biodiversity and sustainability, and forest economics.

Columbia Journal of Environmental Law. Columbia University Press.

The *Columbia Journal of Environmental Law* is the second-oldest environmental law journal in the nation and is widely regarded as one of the preeminent environmental journals in the country. The journal has more than 600 subscribers including law libraries; law firms; and Federal, local, and state courts. It also has a significant international readership.

Conservation Biology. Blackwell Scientific Publications.

This is an international, interdisciplinary journal focusing on the science and development of conservation biology and its application to modern ecological problems.

Conservation Ecology. The Resilience Alliance.

Conservation Ecology is an electronic, peer-reviewed, multidisciplinary journal devoted to the rapid dissemination of current research. Manuscript submission, peer review, and publication are all handled on the Internet. Content ranges from the applied to the theoretical on topics relating to the ecological, political, and social foundations for sustainable social-ecological systems.

Conservation in Practice. Society for Conservation Biology.

This journal provides information to conservation practitioners and policymakers. Articles and topics include conservation biology and ecosystem management.

Ecology Law Quarterly. University of California, School of Law.

One of the most respected and widely read journals on environmental law and policy, *Ecology Law Quarterly (ELQ)* provides fresh insights and analysis from leading authors on critical environmental affairs. Synthesizing legal and technical matters, *ELQ*'s articles are cited frequently in court opinions, by legal institutions, and by attorneys.

Ecology USA. Business Publishers Inc.

This journal provides comprehensive coverage of the ecosystem as a whole with its complex interrelationships.

Environmental Law. Northwestern School of Law of Lewis and Clark College.

Environmental Law, the nation's oldest law review dedicated solely to environmental issues, is a premier legal forum for environmental and natural resources scholarship. It is published quarterly by the students of Lewis and Clark Law School.

Environmental Law Reporter. Environmental Law Institute.

This is a periodical publication on environmental case law, court rulings, and legal analysis of environmental issues.

Environmental Management. Springer-Verlag.

Environmental Management publishes research and opinions concerning the use and conservation of natural resources, the protection of habitats, and the control of hazards. Its field is applied ecology in the widest sense, without regard to the disciplinary boundaries created by modern academic study. Contributions are drawn from biology, botany, climatology, ecology, fisheries management, forest sciences, geography, geology, information science, law politics, public affairs, zoology, and a wide variety of other disciplines, often in combinations determined by interdisciplinary study.

Forest Ecology and Management. Elsevier Science.

Forest Ecology and Management publishes scientific articles concerned with forest management and conservation, and in particular the application of biological, ecological, and social knowledge to the management of planted and natural forests. The journal aims to encourage communication between scientists in disparate fields who share a common interest in ecology and natural resource management and to bridge the gap between research workers and forest managers in the field to the benefit of both.

Forest Policy and Economics. Elsevier Science.

Forest Policy and Economics is an international journal dealing with policy issues, including economics and planning, relating to the forest and forest industries sector. Its aims are to publish original papers of a high scientific standard and to enhance communications among researchers, legislators, decision-makers, and other professionals concerned with formulating and implementing policies.

Forest Products Journal. Forest Products Research Society.

Forest Products Journal is the source of information for industry leaders, researchers, teachers, students, and anyone interested in today's forest products industry.

Forest Science. Society of American Foresters.

Forest Science is internationally renowned as a leading forestry research journal. For more than 30 years, it has been publishing significant articles in forestry research: silviculture, soils, biometry, disease, recreation, photosynthesis, and tree physiology as well as all aspects of management, harvesting, and policy analysis. *Forest Science* also features reviews of recent publications.

Forestry. Oxford Press.

Forestry publishes refereed papers covering the results of original research and practice in forestry. Topics covered are the basic sciences (for example, forest physiology, ecology, classical and molecular genetics, soils, mycology, zoology, economics, and wood structure) and the application of these sciences to the sustainable management of forests.

Forestry Chronicle. Canadian Institute of Forestry.

The *Forestry Chronicle* provides information about scientific management of forests and their resources to forest professionals and practitioners in Canada and around the world, giving them a means to communicate with their peers in the professional community.

Forestry Source. Society of American Foresters.

The *Forestry Source,* the newspaper of the Society of American Foresters, offers timely news for forest resource professionals. It is published 12 times a year and covers the latest forestry policy issues, developments in forestry research and technology, Society of American Foresters programs and activities, and much more.

Harvard Environmental Law Review. Harvard University Press.

The *Harvard Environmental Law Review (HELR)* publishes articles about a broad range of environmental affairs such as land use; air, water, and noise regulation; toxic substances control; radiation control; energy use; workspace pollution; science and technology control; and resource use and regulation. *HELR* draws on the expertise of environmental experts from government, academia, private practice, industry, and public interest groups to cover legal developments from the local to international levels.

In Practice. Institute of Ecology and Environmental Management.

This journal contains news and articles of interest to professional ecologists. It offers vocational advice and covers professional standards.

Institute for Social Ecology Newsletter. Institute for Social Ecology.

This newsletter covers anarchism, community development, political and social change, ecological design, ecological technology, and biological agriculture.

International Forestry Review. Commonwealth Forestry Association.

The *International Forestry Review* is a peer-reviewed scientific journal that publishes papers, research notes, and book reviews on all aspects of forestry and forest research. It is published four times per year. Theme editions are a regular feature and attract a wide audience.

International Journal of Forest Engineering. University of New Brunswick.

The *International Journal of Forest Engineering* is dedicated to the dissemination of scholarly writings on all aspects of forest operations. The scope normally covered includes tree harvesting, processing, and transportation; stand establishment, protection, and tending; operations planning and control; machine design, management, and evaluation; forest access planning and construction; human factors engineering; and education and training.

International Journal of Wildland Fire. Commonwealth Scientific and Industrial Research Organisation (CSIRO) Publishing.

International Journal of Wildland Fire publishes new and significant papers that advance basic and applied research concerning wildland fire. The journal has an international perspective, because wildland fire plays a major social, economic and ecological role around the world.

Journal of Ecology. Blackwell Publishers.

Journal of Ecology is published six times a year and includes original research papers on all aspects of the ecology of plants (including algae) in both aquatic and terrestrial ecosystems. Editors are happy to consider studies of plant communities; populations or individual species; and studies of the interactions between plants and organisms such as animals, fungi, or bacteria, providing that the focus is toward the ecology of the plants.

Journal of Environmental Management. Elsevier Science.

The *Journal of Environmental Management* publishes papers on all aspects of management and use of the environment, both natural and artificial. It is aimed not only at the environmental manager but at everyone concerned with the wise use of environmental resources. The journal tries particularly to publish examples of the use of modern mathematical and computer techniques and encourages contributions from the developing countries.

Journal of Forestry. Society of American Foresters.

The *Journal of Forestry* is the premier scholarly journal in forestry. It has received several national awards for excellence. The mission of the journal is to advance the profession of forestry by keeping professionals informed about significant developments and ideas in forest science, natural resources management, and forest policy.

Journal of Land, Resources, and Environmental Law. University of Utah, College of Law.

This journal publishes traditional legal articles as well as those of an interdisciplinary nature. It also includes expository articles that provide the nonspecialist with an overview of specialized areas of law, and it publishes articles oriented toward a practitioner's perspective.

Journal of Natural Resources and Environmental Law. University of Kentucky, College of Law.

This law journal focuses on the legal, policy, and ethical issues relating to environmental and natural resources law.

Journal of Nature Conservation. European Center for Nature Conservation.

This is an international journal devoted to nature, natural resource conservation, and environment.

Journal of Pulp and Paper Science. Pulp and Paper Technical Association of Canada and Technical Association for Pulp and Paper Industry (TAPPI).

The *JPPS* is devoted to the science of pulp and paper, and its aim is to publish articles that illuminate the underlying principles of the technology of pulp and paper rather than those that are of a purely technological or engineering nature.

Journal of Sustainable Forestry. Haworth Press.

This journal provides a linkage of silviculture and the underlying biology: tree physiology, morphology, and genetics. As such, it elucidates the scientific principles and techniques of controlling, protecting, and restoring the regeneration, composition, and growth of natural forest vegetation as well as plantations, agroforestry, and silvo-pastoral systems. It encompasses topics from biotechnology, physiology, silviculture, wood science, economics, and forest management.

Journal of Tropical Forestry. Scientific Publishers.

An international, interdisciplinary journal focusing on the science, management, and conservation of tropical forests and forest lands around the globe.

Land Letter: The Newsletter for Natural Resource Professionals. Environmental and Energy Publishing LLC.

This newsletter reports on natural resources issues and environmental studies.

Landscape Ecology. Kluwer Academic Publishers.

Landscape Ecology seeks new and innovative papers that improve understanding of the relationship between patterns and processes, explaining the spatial variation in landscapes at multiple scales as affected by natural causes and human society. Because landscape ecology is a broad, interdisciplinary topic, the journal is open to contributions that consider the landscape as the basis of integration of knowledge.

National Parks. National Parks and Conservation Association.

This is the only national publication focusing solely on National Parks. The most important communication vehicle of the National Parks and Conservation Association, the magazine creates an awareness of the need to protect and properly manage the resources found within and adjacent to the National Parks.

Natural Resource Perspectives. Overseas Development Institute.

This journal presents accessible information on international development issues.

Natural Resources and Environment, in *The Section of Natural Resources Law.* American Bar Association.

Natural Resources and Environment is a quarterly magazine containing practical, informative articles for legal practitioners. The section provides a wide range of publications to enhance the practices of environmental, energy, and resources lawyers.

Natural Resources and Environmental Issues. Utah State University, College of Natural Resources.

This journal publishes monographs devoted primarily to research in natural resources and related fields.

Natural Resources Forum. Blackwell Publishing Ltd.

This journal focuses on sustainable development and management of natural resources, mainly in developing countries. In recent years, following the growing importance of social policy and programs at the United Nations, the journal's editorial policy has also shifted to socioeconomic aspects of sustainable development of water, energy, and mineral resources.

Natural Resources Management. Australian Association of Natural Resource Management.

This journal features technical and research articles that cover all aspects of natural resource management in Australia.

Nature and Resources. United Nations Educational, Scientific, and Cultural Organization (UNESCO).

This publication is a production of UNESCO and focuses on the status of the world's environment and natural resources and their relationship with global society and economy.

Naturopa. European Information Center for Nature Conservation.

The magazine *Naturopa*, which covers specific topics in each issue, is published three times a year in five languages (English, French, Italian, German, and Russian). From 1968 to 2000 *Naturopa* concentrated on promoting nature conservation, sustainable management of natural resources, and the development of a multidisciplinary approach to environmental issues. In 2001, *Naturopa* began the progressive introduction of new themes such as cultural heritage and landscape preservation in a perspective of sustainable development and enhancement of the quality of life.

OnEarth, formerly *The Amicus Journal.* Natural Resources Defense Council.

OnEarth is an independent quarterly magazine of environmental thought and opinion. It publishes items from hard-hitting reports on the environmental issues that define our time—global warming, "big oil" in the Arctic, the disappearance of wild places—to short takes on the new, the provocative, and the just plain weird.

Quarterly Journal of Forestry. Hall-McCartney Ltd.

This journal is a periodical of the Royal Forestry Society of England, Wales, and Northern Ireland, providing applied forestry techniques, technical features, and book reviews.

Regional Journals of Applied Forestry. Society of American Foresters.

These journals focus on research, practice, and techniques targeted to foresters and allied professionals in specific regions of the United States and Canada. The regional journals are the *Northern Journal of Applied Forestry,* the *Southern Journal of Applied Forestry,* and the *Western Journal of Applied Forestry.*

Renewable Resources Journal. Renewable Natural Resources Foundation.

This journal is a forum for discourse on public policy issues. Cooperation and understanding among professionals in the various natural resources disciplines can be enhanced by sharing perspectives on resource issues, objectives, and technical and political constraints. To this end, the journal carries original articles, previously unpublished speeches, and selected articles from the publications of member organizations. Communications among natural resource professionals also is promoted by publication of news, announcements, and meeting notices of member organizations and other groups.

Sierra. The Sierra Club.

Sierra is the official magazine of the Sierra Club.

Society and Natural Resources. Taylor and Francis.

Bringing together social science research on current and emerging environmental and natural resource issues, *Society and Natural Resources* provides a forum for scientific, refereed research that underlies management decisions about natural resource development from multidisciplinary and interdisciplinary social science perspectives.

Timber Harvesting. Hatton-Brown Publishers.

Timber Harvesting appeals to business owners and operating management of the U.S. logging and forestry industry. *Timber Harvesting* focuses on operational/technological developments in logging, timber transportation, and silvicultural operations. It is the only logging magazine that routinely surveys its subscribers about equipment purchasing and harvesting practices and is the leader in reporting up-to-date environmental developments affecting the industry.

Tulane Environmental Law Journal. Tulane Law School, Tulane Environmental Law Society.

The *Tulane Environmental Law Journal* is a student-run and student-edited law review devoted to environmental issues. It is published twice each year and has a national circulation. The journal publishes articles by scholars and practitioners as well as student work.

Unasylva. United Nations Food and Agriculture Organization (FAO).

Unasylva, an international journal of forestry and forest industries, is produced quarterly in separate English, French, and Spanish editions. *Unasylva* covers all aspects of forestry: policy and planning; conservation and management of forest-based plants and animals; rural socioeconomic development; species improvement; industrial development; international trade; and environmental considerations, including the role of forests and trees in maintaining a sustainable base for agricultural production at the micro and macro levels as well as the effects of environmental change on forestry.

Urban Ecology. Urban Ecology Inc.

This journal discusses issues of urban planning and design in the context of social justice. It covers such ecological issues as mass transportation, urban forests, and aesthetic improvement.

Urban Forestry and Urban Greening. Urban & Fischer Verlag.

This publication concentrates on all tree-dominated as well as other green resources in and around urban areas, such as woodlands, public and private urban parks and gardens, urban nature areas, street tree and square plantations, botanical gardens, and cemeteries.

Virginia Environmental Law Journal. University of Virginia, School of Law.

The *Virginia Environmental Law Journal,* a leader in environmental legal scholarship, is published by students at the University of Virginia School of Law. The journal focuses on state, regional, national, and international issues of environmental, preservation, land use, energy, and natural resources law.

Wilderness. The Wilderness Society.

This is the official monthly publication of the Wilderness Society.

Women in Natural Resources. University of Idaho, Bowers Laboratory.

Women in Natural Resources is a journal of professional women in forestry, wildlife, fisheries, range, soils, and the social sciences as they pertain to natural resources.

Wood and Fiber Science. Allen Publishing Inc.

Wood and Fiber Science, the official publication of the Society of Wood Science and Technology, publishes papers with both professional and technical content. It considers original papers of professional concern or those based on research dealing with the science, processing, and manufacture of wood and composite products of wood or wood fiber origin.

Woodland Report: Late-Breaking News on Private Forestry Issues from Washington D.C. and State Capitals. National Woodland Owners Association.

The report contains practical forestry information for private woodland owners.

World Conservation. International Union for Conservation of Nature and Natural Resources (IUCN).

IUCN's *World Conservation* journal is published three times a year in English, French, and Spanish. Each issue contains articles focusing on various conservation themes, including mountains, threatened species, world heritage, and forests. The articles solicited from guest contributors, scientists, and experts provide a wide range of viewpoints on what is involved in preserving natural resources and supporting sustainable development. The journal also includes a section dedicated to news and current activities in IUCN and its commissions.

World Forestry Directions. Sterling Publications Ltd.

This journal provides a comprehensive review of the emerging technologies that will drive the forestry industry in the future.

8

Nonprint Resources

Numerous sources of information on forests, forestry, and forest conservation can be found on the World Wide Web and on videotape. Besides providing information about government and nongovernment organizations (see chapter 6, U.S. Forestry Organizations), many web sites serve as online information clearinghouses, providing links to a vast array of more specific online resources.

With the relatively low cost of publishing online, large amounts of data, information, maps, and reports are now accessible. The availability of information is staggering, and one can learn a lot about forests, forestry, and forest conservation policy from surveying a broad array of web sites. However, data, information, maps, and analyses posted on the World Wide Web often do not undergo rigorous review and critique; users should always cross-check information and analyses with sources that have undergone more critical review, especially consulting peer-reviewed research when possible.

Videos are also available through various specialty media organizations. Videos are especially useful for educational purposes and broad audiences.

Information Clearinghouses on the Web

Environmental News Service
http://ens-news.com/

Daily news stories and analysis on a broad range of environmental issues globally.

Environmental Organization Web Directory
http://webdirectory.com/

A clearinghouse of numerous government, nongovernment, community, and educational organizations under 30 topic headings.

FireWise
http://www.firewise.org/

Information and resources aimed at homeowners and property owners in the wildland-urban interface for mitigating risks from wildland forest fires.

Forest Conservation Portal
http://forests.org/

Forests.org Inc., which operates this web site, works to end deforestation, preserve old-growth forests, conserve and sustainably manage other forests, maintain climatic systems, and commence the age of ecological restoration. This web site is for educational and noncommercial use only.

ForestInformation.com
http://www.forestinformation.com/index.asp

Web-based clearinghouse of forest and forestry information pertaining to the United States.

Inventory and Monitoring Institute
http://www.fs.fed.us/institute/

Administered by the Forest Service of the U.S. Department of Agriculture, the Inventory and Monitoring Institute provides technical consultation to Forest Service units with responsibilities for on-the-ground inventory, monitoring, and planning activities. The Inventory and Monitoring Institute's work focuses on the application of knowledge and technology to these areas of the information environment:

- Data collection with sound inventory design and quality assurance;
- Land classification using Bailey's world-class ecoregion principles;

- Information management using leading-edge Forest Service information technology;
- Information analysis to answer questions and address issues;
- Knowledge sharing through technical assistance to other countries.

National Interagency Fire Center
http://www.nifc.gov/

Provides up-to-date data and information about wildland fires throughout the United States as well as links to wildland fire organizations, training opportunities, statistics of previous fire seasons, and the latest fire science and technology.

Timbergreen Forestry
http://my.execpc.com/~tmbrgrn/index.html

Timbergreen Forestry works with other forest owners to grow large, high-quality crop trees that give many benefits from careful management. The web site is directed at SmartWood–certified resource managers to sustain profitability of small private forest lands.

Vegetation Management Tools
http://www.fs.fed.us/vegtools/index.shtml

Web-based information-sharing source for projects used in the field to restore and maintain forest health.

Online Reports, Maps, and Information

Biodiversity and Sustainable Forestry: State of the Science Review. National Commission on Science for Sustainable Forestry.
http://cnie.org/NCSSF/Documents/Biodiversity%20Paper.logo1.pdf

This paper focuses on summarizing the research relevant to the current status of science and research efforts related to sustainable forestry practices and biodiversity.

Descriptions of the Ecoregions of the United States. U.S. Department of Agriculture, U.S. Forest Service.
http://www.fs.fed.us/land/ecosysmgmt/ecoreg1_home.html
http://www.neartica.com/ecology/ecoreg/ecoprov.
htm#anchor170336

Digitized maps from Robert G. Bailey's "Ecoregions of the United States."

Digital Representations of Tree Species Range Maps of the United States. U.S. Department of the Interior, U.S. Geological Survey (USGS).
http://climchange.cr.usgs.gov/data/atlas/little/

Maps of the ranges of tree species in North America—compiled by Elbert Little of the Forest Service and others—were digitized for use in USGS's vegetation-climate modeling studies. These digital map files are available here for download.

Ecoregions of the United States. U.S. Environmental Protection Agency (EPA).
http://www.epa.gov/bioindicators/html/usecoregions.html

Digitized maps of ecoregions of the United States as interpreted by the EPA.

Forest Land Distribution Data for the United States. U.S. Department of Agriculture, U.S. Forest Service.
http://www.srsfia.usfs.msstate.edu/rpa/rpa93.htm

Downloadable maps and metadata files on forest land data for the United States.

Global Forest Resources Assessment 2000: Main Report. Food and Agriculture Organization.
http://www.fao.org/forestry/fo/fra/main/index.jsp

The "Global Forest Resources Assessment 2000" is an appraisal of the status of the world's forests, with analyses of changes since 1980. The assessment is a key reference source for forest information for governments, nongovernmental organizations, and international conventions such as the Convention on Biological Diversity and the United Nations Framework Convention on Climate Change.

Hayman Fire Case Study Analysis. U.S. Department of Agriculture (USDA), U.S. Forest Service.
http://www.fs.fed.us/rm/hayman_fire/

A report compiled by the Hayman fire case study analysis team comprising researchers from the USDA Forest Service Rocky Mountain Research Station in cooperation with USDA Forest Service Rocky Mountain Region and the Colorado Forest Service. This team of federal, state, and local experts from throughout the United States developed an analysis framework to address issues raised by Congressman Mark Udall (D-Colorado) in connection with the Hayman fire, which burned 138,000 acres in Colorado in 2002. Teams addressed fire behavior, home destruction, social and economic impacts, fire rehabilitation, and ecological effects. Using Representative Udall's issues, each team developed a set of analysis questions and its own approach to answering the questions in a timely manner. Techniques used by the teams included interviews, analysis of existing data, expert opinion, Hayman fire reports, and other available information. The report highlights each team's interim findings addressing the analysis questions.

Temperate and Boreal Forest Resources Assessment 2000. Food and Agriculture Organization (FAO).
http://www.unece.org/trade/timber/fra/welcome.htm

The United Nations Economic Commission for Europe (UNECE)/FAO published and distributed a CD-ROM containing the "Temperate and Boreal Forest Resources Assessment 2000" (TBFRA-2000) electronic database and the main TBFRA-2000 report. Since the launch of the CD-ROM, positive responses from users have confirmed that this product is another important source of information about temperate/boreal forests and a useful tool for scientists, researchers, and all others who are interested in the TBFRA data.

Understanding the FAO Forest Resources Assessment 2000. World Resources Institute.
http://www.wri.org/press/fao_fra5.html

Critique of the methods and findings of the "Global Forest Resources Assessment 2000: Main Report" by the Food and Agriculture Organization.

United Nations Framework Convention on Climate Change. United Nations.
http://unfccc.int/cop4/conv/conv.html

Online summary and full report of the Kyoto Protocol.

Videos

Can Tropical Forests Be Saved?
Type: VHS videocassette
Length: 120 minutes
Date: 1991
Cost: Academic institutions $240.00; community groups and libraries $150.00
Source: Richter Productions
http://www.richtervideos.com

Fifty years ago, tropical rainforests covered 14 percent of the planet's surface in a belt of 33 countries, most of them grouped along the equator. More than half of this forest has now been lost; much of it is now barren wasteland. Explore alternative approaches to conservation, including debt-for-nature swap and sustainable harvesting.

Earth Report
Type: VHS videocassette
Length: 250 minutes on two tapes; 130 minutes on five tapes
Date: 1996
Source: Television Trust for the Environment (TVE)
Prince Albert Road
London NW1 4RZ
United Kingdom
http://www.tve.org/network.html

A regular update on the state of the planet, these tapes are a report card on how well the world is achieving the targets of the 1992 Earth Summit in Rio de Janeiro, Brazil. It breaks the mold of doom-and-gloom coverage of the environmental crisis and documents how people around the globe are rising to their environmental challenges.

Fate of the Forest
Type: VHS videocassette

Length: 30 minutes
Date: 1996
Source: Television Trust for the Environment (TVE)
Prince Albert Road
London NW1 4RZ
United Kingdom
http://www.tve.org/network.html

For more than 20 years, the shrinkage of the world's tropical forests has been at the forefront of environmental concerns. Find out how local people ingeniously imitate natural productivity, agroforestry techniques, and other initiatives.

The Salmon Forest
Type: VHS videocassette
Video Number: 1-56029-909-6
Length: 52 minutes
Date: 2001
Cost: Purchase $250.00; rental $75.00
Source: Bullfrog Films
372 Dautrich Road
Reading, PA 19606
http://www.bullfrogfilms.com

Reveals the fragile connections among salmon, bears, trees, and people in the Pacific Northwest rainforest.

The Temperate Rainforest
Type: VHS videocassette
Video Number: 0-7722-0394-6
Length: 16 minutes
Date: 1985
Cost: Purchase $49.00; rental $20.00
Source: Bullfrog Films
372 Dautrich Road
Reading, PA 19606
http://www.bullfrogfilms.com

This beautifully photographed film examines the characteristics and ecology of the coastal rainforest of the Pacific Northwest.

Trees Are the Answer
Type: VHS videocassette
Length: 30 minutes

Date: 2000
Cost: $19.95
Source: Greenspirit Ltd.
4068 West 32nd Avenue
Vancouver, BC V6S 1Z6
Canada
http://www. greenspirit.com/index.cfm

Trees Are the Answer is an enlightening video documentary hosted by Greenpeace cofounder Dr. Patrick Moore. It explores the beauty, biodiversity, and resiliency of forests while delivering a powerful message about how forests work and the role they play in solving many current environmental problems.

Tree-Sit: The Art of Resistance
Type: VHS videocassette
Length: 120 minutes
Date: 1998
Cost: Institutions $200.00; individuals $25.00
Source: Earth Films/ESP
P.O. Box 2198
Redway, CA 95560
http://www.earthfilms.org/

Feature-length grassroots documentary. Covers the struggle to save Headwaters Forest, the attempted assassination of Judi Bari, the pepper-spray torture of young activists, and the establishment of permanent "tree villages" hundreds of feet up; culminates in the World Trade Organization protests in the streets of Seattle. This film is historical, exhilarating, informative, and intense!

Up in Flames: A History of Fire Fighting in the Forest
Type: VHS videocassette
Length: 29 minutes
Cost: $25.00
Source: Forest History Society
701 William Vickers Avenue
Durham, NC 27701
(919) 682-9319
http://www.lib.duke.edu/forest/Publications/films.html

Up in Flames documents the development of North American fire detection, communication, and fire suppression technology from the inception of the lookout tower to the weather satellite.

Nonprint Resources from Various States

Listed are a few states that have fairly extensive online and video resources.

The Forestry Media Center in the College of Forestry at Oregon State University
http://www.cof.orst.edu/cof/fmc/

Oregon State University has a large collection of educational, informational, and technical videos on various forestry topics. To download an order form from the Internet, go to http://www.cof.orst.edu/cof/fmc/orderpage.php.

Conversations on Sustainable Forestry
Type: VHS videocassette
Length: 90 minutes
Date: 1991
Number: 992
Cost: $95.00; rental $25.00

Forest Fragmentation: Issues and Implications
Type: VHS videocassette
Length: 18 minutes
Date: 1990
Number: 968
Cost: $95.00; rental $25.00

Help for Your Woodlands
Type: VHS videocassette
Length: 20 minutes
Date: 1991
Number: 971
Cost: $130.00; rental $25.00

The Huckleberry Story: Building a Bridge between Culture and Science
Type: VHS videocassette
Length: 20 minutes
Date: 1997
Number: 1097
Cost: $95.00; rental $25.00

Management Planning for Your Small Woodland: An Introduction
Type: VHS videocassette
Length: 19 minutes

Date: 1996
Number: 1017
Cost: $95.00; rental $25.00

Managing for Biodiversity in Young Forests
Type: VHS videocassette
Length: 28 minutes
Date: 2000
Number: 1155
Cost: $95.00; rental $25.00

Perspectives on Ecosystem Management
Type: VHS videocassette
Length: 145 minutes
Date: 1988
Number: 941
Cost: $95.00; rental $25.00

Thinning Young Stands
Type: VHS videocassette
Length: 31 minutes
Date: 1998
Number: 1089
Cost: $95.00; rental $25.00

Maryland Cooperative Extension

The Maryland Cooperative Extension Service offers a video about riparian forest buffers. To order, contact Robert Tjaden, Wye Research and Education Center, P.O. Box 169, Queenstown, MD 21658, (410) 827-8056.

Riparian Forest Buffers: The Link between Land and Water
Type: VHS videocassette

Oklahoma Cooperative Extension

To order, write to Oklahoma State University Agricultural Communications, 111 PIO Building, Oklahoma State University, Stillwater, OK 74078-6041.

Logging, Best Management Practices and Water Quality
Type: VHS videocassette
Length: 21 minutes
Number: VT 264
Cost: $24.95

Managing Forest Ecosystems: Assessing New Opportunities
Type: VHS videocassette
Length: 105 minutes
Number: TC 146
Cost: $24.95

Pennsylvania State University Cooperative Extension Service, College of Agricultural Sciences

To order publications and videos, please print and fill out an order form available at the following web site: http://pubs.cas. psu.edu/PubCat.OrderForm.pdf. Mail it to the Publications Distribution Center, The Pennsylvania State University, 112 Agricultural Administration Building, University Park, PA 16802-2602, (814) 865-6713, Fax: (814) 863-5560.

Forest Stewardship at the Urban/Rural Interface
Type: VHS videocassette
Length: 27 minutes
Date: 1992
Number: L36540VH
Cost: $35.00

Glossary

Acre An area of land measuring about 43,560 square feet. A square 1-acre plot measures about 209 feet by 209 feet.

Adaptive management A management philosophy that rigorously combines management, research, monitoring, and means of changing practices so that credible information is gained and management activities are modified by experience.

Afforestation Establishment of forest by artificial methods, such as planting or sowing on land where trees have never grown.

Age class One of the intervals, commonly 10 or 20 years, into which the age range of tree crops is divided for classification or use. Also pertains to the trees included in such an interval.

All-aged or uneven-aged management The practice of managing a forest by periodically selecting and harvesting individual trees or groups of trees from the stand while preserving its natural appearance. Most common in hardwood forests.

All-aged or uneven-aged stand A forest stand composed of trees of different ages and sizes.

Allowable cut Volume of timber that may be harvested during a given period to maintain sustained production.

Allowable cut effect Allocation of anticipated future forest timber yields to the present allowable cut; this is employed to increase current harvest levels by spreading anticipated future growth over all the years in the rotation.

Anadromous fish Fish that breed in fresh water but live their adult life in the sea. On the Pacific coast, anadramous fish include all the Pacific salmon, steelhead trout, some cutthroat trout, and Dolly Varden char, lampreys, and eulachons.

Animal unit month (AUM) The amount of forage required for one month by an average cow; unit used to calculate grazing management.

Annual allowable harvest   Quantity of timber scheduled to be removed from a particular management unit in one year.

Annual growth   Average annual increase in the biomass of growing-stock trees of a specified area.

Aquatic habitat   Habitat where a variety of marine or freshwater flora and fauna occur for long periods throughout the year.

Area regulation   Method of controlling the annual or periodic acreage harvested from a forest, despite fluctuations in fiber-yield volumes. Leads to a managed forest.

Artificial regeneration   Renewal of the forest by planting or direct seeding; establishing a new stand of trees by planting seeds or seedlings by hand or machine.

Backfire   Blaze set in front of an advancing forest fire in an effort to check the wildfire by cutting off its fuel supply.

Basal area   The cross-section area of a tree stem in square feet commonly measured at diameter breast height (4.5 feet above ground) and inclusive of bark.

Benefit/cost analysis   Technique for comparing alternate courses of action by an assessment of their direct and indirect outputs (benefits) and inputs (costs). Benefits and costs are usually defined in economic and social terms.

Best management practices (BMP)   Standards and guidelines set by professionals that define the most appropriate means and mechanisms for forest practices.

Biological diversity   The variety of life forms in a given area. Diversity can be categorized in terms of the number of species, the variety in the area's plant and animal communities, the genetic variability of the animals, or a combination of these elements.

Biomass   Total woody material in a forest. Refers to both merchantable material and material left following a conventional logging operation. In the broad sense, all of the organic material on a given area; in the narrow sense, burnable vegetation to be used for fuel in a combustion system.

Board foot   A unit of wood measuring 144 cubic inches. A 1-inch by 12-inch shelving board that is 1 foot long is equal to 1 board foot. Board foot volume is determined by: length (feet) × width (inches) × thickness (inches)/divided by 12.

Broadcast burning   Controlled burn, in which the fire is intentionally ignited and allowed to proceed over a designated area within well-defined boundaries, for the reduction of fuel hazard after logging or for site preparation before planting. Also called slash burning.

Broadleaf   A tree with leaves that are broad, flat, and thin and generally shed annually.

Buffer strip   A narrow zone or strip of land, trees, or vegetation bordering an area. Common examples include visual buffers, which screen the view along roads, and streamside buffers, which are used to protect water quality. Buffers may also be used to prevent the spread of forest pests.

Carrying capacity   The average number of individuals that can be sustained on a unit of land, compatible with management objectives for the unit. It is a function of site characteristics, management goals, and management intensity.

Chip   Small piece of wood used to make pulp. Chips are made either from wood waste in a sawmill or pulpwood operation, or from pulpwood specifically cut for this purpose. Chips are larger and coarser than sawdust.

Clearcut harvest   A harvesting and regeneration method that removes all trees within a given area. Clearcutting is most commonly used in pine and hardwood forests, which require full sunlight to regenerate and grow efficiently.

Commercial thinning   Partial harvesting of a stand of trees for economic gains from the harvested trees and to accelerate the growth of the trees left standing.

Community forestry   Forest planning, management, and monitoring by and for local communities.

Conifer   Tree that is usually evergreen with cones and needle-shaped or scalelike leaves, producing wood known commercially as softwood.

Conservation   Management of the human use of the biosphere so that it may yield the greatest sustainable benefit to present generations while maintaining its potential to meet the needs and aspirations of future generations. It includes the preservation, maintenance, sustainable utilization, restoration, and enhancement of the environment.

Conservation biology   The discipline that treats the content of biodiversity, the natural processes that produce it, and the techniques used to sustain it in the face of human-caused environmental disturbance.

Conservation easements   A land protection tool to remove private land from future development, often in the form of purchasing development rights on private property.

Controlled burning   Use of fire to destroy logging debris, reduce buildups of dead and fallen timber that pose wildfire hazards, control tree diseases, and clear land. Other functions of a controlled burn include clearing a buffer strip in the path of a wildfire; see *Backfire*.

Cord   A stack of round or split wood consisting of 128 cubic feet of wood, bark, and air space. A standard cord measures 4 feet by 4 feet by 8 feet. A face cord or short cord is 4 feet by 8 feet by any length of wood under 4 feet.

Cost-share assistance An assistance program offered by various state and Federal agencies that pays a fixed rate or percentage of the total cost necessary to implement some forestry or agricultural practice.

Critical habitat Those areas officially designated by the Secretary of Interior or Secretary of Commerce as needed for survival and recovery of listed species under the Endangered Species Act of 1973.

Cruise A survey of forest land to locate timber and estimate its quantity by species, products, size, quality, or other characteristics.

Cumulative effects Effects on biota of stress imposed by more than one mechanism (e.g., stress in fish imposed by both elevated suspended sediments concentrations in the water and by high water temperature).

Cutting cycle The planned time interval between major harvesting operations within the same stand—usually within uneven-aged stands. For example, on a 10-year cutting cycle in a hardwood stand, trees are harvested every 10 years.

DBH See *Diameter at breast height.*

Deciduous Perennial plants that are normally leafless for some time during the year.

Defoliators Insects that destroy foliage.

Dendrology Study and identification of trees.

Diameter at breast height (DBH) Tree DBH is outside bark diameter at breast height. Breast height is defined as 4.5 feet (1.37 m) above the forest floor on the uphill side of the tree. For the purposes of determining breast height, the forest floor includes the duff layer that may be present but does not include unincorporated woody debris that may rise above the ground line.

Diameter classes Classification of trees based on diameter outside bark measured at DBH. In forest surveys, each diameter class encompasses approximately 2 inches: the 6-inch class would include trees 5.0 through 6.9 inches in DBH.

Diameter limit Maximum diameter of trees to be cut, as in a timber sales contract.

Ecological health Both the occurrence of certain attributes that are deemed to be present in a healthy, sustainable resource, and the absence of conditions that result from known stresses or problems affecting the resource.

Ecological integrity The quality of a natural unmanaged or managed ecosystem in which the natural ecological processes are sustained, with genetic, species, and ecosystem diversity assured for the future.

Ecological type (also see *Habitat*) A category of land having a unique combination of potential natural community in terms of soil, landscape features, and climate, and differing from other ecological types in its ability to produce vegetation and respond to management. Classes of ecological types include all sites that have this unique combination of components with the defined range of properties.

Ecology   The science or study of the relationships between organisms and their environment.

Ecosystem   A functional unit consisting of all the living organisms (plants, animals, and microbes) in a given area and all the nonliving physical and chemical factors of their environment, linked together through nutrient cycling and energy flow. An ecosystem can be of any size—a log, pond, field, forest, or the Earth's biosphere—but it always functions as a whole unit. Ecosystems are commonly described according to the major type of vegetation—for example, forest ecosystem, old-growth ecosystem, or range ecosystem.

Edge   The transition between two different types or ages of vegetation.

Endangered or threatened species   A species is endangered when the total number of remaining members may not be sufficient to reproduce enough offspring to ensure survival of the species. A threatened species exhibits declining or dangerously low populations but still has enough members to maintain or increase numbers.

Environment   The interaction of climate, soil, topography, and other plants and animals in any given area. An organism's environment influences its form, behavior, and survival.

Erosion   The wearing away of land or soil by the action of wind, water, or ice.

Even-aged forest   Theoretically, stands in which all the trees are one age. In actual practice, these stands are marked by an even canopy of uniform height characterized by intimate competition between trees of approximately the same size.

Even-aged management   A forest management method in which all trees in an area are harvested at one time or in several cuttings over a short time to produce stands that are all the same age or nearly so. This management method is commonly applied to shade-intolerant conifers and hardwoods.

Fire danger   Measure of the likelihood of a forest fire, based on temperature, relative humidity, wind force and direction, and the dryness of the woods.

Fire hazard   Condition of fuel on the ground, particularly slash.

Fire suppression   All activities concerned with controlling and extinguishing a fire following its detection. Synonymous with fire control.

Firebreak   Any nonflammable barrier used to slow or stop fires. Several types of firebreaks are mineral soil barriers; barriers of green, slow-burning vegetation; and mechanically cleared areas.

Forage   Grasses, herbs, and small shrubs that can be used as feed for livestock or wildlife.

Forest   Area managed for the production of timber and other forest products or maintained as wood vegetation for such indirect benefits as protection of watersheds or recreation.

Forest certification   A labeling process by which forests are determined, through third-party auditing, to be managed under sustainable forest management principles and practices. Forest certification is a means by which end-product consumers can identify if their purchases are derived from sustainably-managed forest land.

Forest cover type   A descriptive term used to group stands of similar characteristics and species composition (due to given ecological factors) by which they may be differentiated from other groups of stands.

Forest health   Forest condition that is naturally resilient to damage; characterized by biodiversity, it contains sustained habitat for timber, fish, wildlife, and humans, and meets present and future resource management objectives.

Forest land   Land at least 10 percent stocked by trees of any size or formerly having had such tree cover and not currently built up or developed for agricultural use. Forest land may include grassland, shrub land, tree land, wetland, and/or barren land. Examples of forest land uses are grazing, recreation, and timber production.

Forest management   The practical application of scientific, economic, and social principles to the administration and working of a forest for specified objectives. Particularly, that branch of forestry concerned with the overall administrative, economic, legal, and social aspects and with essentially scientific and technical aspects—especially silviculture, protection, and forest regulation.

Forest management plan   Written guidelines for current and future management practices recommended to meet an owner's objectives.

Forest practice   Any activity that enhances and/or recovers forest growth or harvest yield, such as site preparation, planting, thinning, fertilization, and harvesting.

Forest resources   Resources and values associated with forests and range including timber, water, wildlife, fisheries, recreation, botanical forest products, forage, and biological diversity.

Forest stewardship plan   A written document listing activities that enhance or improve forest resources (wildlife, timber, soil, water, recreation, and aesthetics) on private land over a five-year period.

Forest type   Classification of forest land in terms of potential cubic-foot volume growth per acre at the culmination of mean annual increment in fully stocked natural stands. Also, classification of forest land based on the species forming a plurality of live-tree stocking. Type is determined on the basis of species plurality of all live trees that contribute to stocking.

Forestry The science, art, and practice of managing and using trees, forests, and their associated resources for human benefit.

Fragmentation The process of transforming large continuous forest patches into one or more smaller patches surrounded by disturbed areas. This occurs naturally through such agents as fire, landslides, windthrow, and insect attack. In managed forests, timber harvesting and related activities have been the dominant disturbance agents.

Fuel loading A buildup of fuels—especially easily ignited, fast-burning fuels.

Fuel management The planned manipulation and/or reduction of living or dead forest fuels for forest management and other land use objectives (such as hazard reduction, silvicultural purposes, and wildlife habitat improvement) by prescribed fire, mechanical, chemical, or biological means and/or changing stand structure and species composition.

Fuelbreak An existing barrier or change in fuel type (to one that is less flammable than that surrounding it), or a wide strip of land on which the native vegetation has been modified or cleared, that acts as a buffer to fire spread so that fires burning into it can be more readily controlled. Often selected or constructed to protect a high-value area from fire.

Genetic diversity Variation among and within species that is attributable to differences in hereditary material.

Geographic information system A computer system designed to allow users to collect, manage, and analyze large volumes of spatially referenced information and associated attribute data.

Grade Established quality or use classification of timber.

Grassland Areas on which vegetation is dominated by grasses, grasslike plants, forbs, and/or cryptogams (mosses, lichens, and ferns), provided these areas do not qualify as built-up land or cultivated cropland. Examples are tallgrass and shortgrass prairies, meadows, cordgrass marshes, sphagnum moss areas, pasturelands, and areas cut for hay.

Green strip Uncut strip of timber left along streams and roads. Also known as buffer strip, leave strip, streamside management zone.

Group selection The selective removal of small groups of trees to regenerate shade-intolerant trees in the opening (usually at least $1/4$ acre).

Growing stock The sum of all trees in a forest or specified part of it.

Habitat (a) An area in which a specific plant or animal can naturally live, grow, and reproduce. (b) For wildlife, habitat is the combination of food, water, cover, and space.

Hardwoods (deciduous trees) Trees with broad, flat leaves as opposed to coniferous or needled trees. Wood hardness varies among the hardwood species, and some are actually softer than some softwoods.

Harvest schedule A document listing the stands to be harvested during a year or period, usually showing types and intensities of harvests for each stand, as well as a timetable for regenerating currently nonproductive areas.

High-grading A harvesting technique that removes only the biggest and most valuable trees from a stand and provides high returns at the expense of future growth potential. Poor-quality, shade-loving trees tend to dominate in these continually high-graded sites.

Historical range of variability The range of the spatial, structural, compositional, and temporal characteristics of ecosystem elements during a period specified to represent "natural" conditions. Also called the *range of natural variation.*

Human dimension An integral component of ecosystem management that recognizes people are part of ecosystems; that people's pursuits of past, present, and future desires, needs, and values (including perceptions, beliefs, attitudes, and behaviors) have influenced and will continue to influence ecosystems; and that ecosystem management must include consideration of the physical, emotional, mental, spiritual, social, cultural, and economic well-being of people and communities.

Hydrology The science that describes and analyzes the occurrence of water in nature and its circulation near the surface of the Earth.

Improvement cut An intermediate cut made to improve the form, quality, health, or wildlife potential of the remaining stand.

Incentive A reward for improving forest management. Incentives include reimbursement of some expenses but can also take the form of an abatement of property or income tax.

Indicator species Species of plants or animals used to predict site quality and characteristics.

Industrial private forest Private forest land owned by an individual or organization that directly supplies a production mill.

Intensive forest management Use of a wide variety of silvicultural practices, such as planting, thinning, fertilization, harvesting, and genetic improvement, to increase the capability of the forest to produce fiber.

Keystone species A species that plays an important ecological role in determining the overall structure and dynamic relationships within a biotic community. Presence of a keystone species is essential to the integrity and stability of a particular ecosystem.

Ladder fuels Fuels that provide vertical continuity between the surface fuels and crown fuels in a forest stand, thus contributing to the ease of torching and crowning.

Land and resource management plan A strategic integrated resource plan at the National Forest level. It is based on the principles of enhanced public involvement, consideration of all resource values, consensus-based decision-making, and resource sustainability.

Landscape The fundamental traits of a specific geographic area, including its biological composition, physical environment, and anthropogenic or social patterns.

Landscape ecology The study of the distribution patterns of communities and ecosystems, the ecological processes that affect those patterns, and changes in pattern and process over time.

Leave strip Strip of uncut timber left between cutting units or adjacent to another resource such as a stream. Also known as a buffer strip, green strip, or streamside management zone.

Log rule or log scale A table based on a diagram or mathematical formula used to estimate volume or product yield from logs and trees. Scribner is the common scale for pine; Doyle is the common hardwood scale; and the International $1/4$ Inch Rule best measures mill output, although it is used less frequently than the other log scales.

Management plan A written document describing the goals, objectives, and proposed actions over a specified time period for a specific unit of land. Contains inventory and other resource data.

Marginal land Land that does not consistently produce a profitable crop because of infertility, drought, or other physical limitations such as shallow soils.

Mature tree A tree that has reached a desired size or age for its intended use. Size, age, or economic maturity varies depending on the species and intended use.

MBF Abbreviation denoting 1,000 board feet. MBF is a typical unit of trade for dimension lumber and sawtimber stumpage. (It takes 11 MBF of wood to build an average 1,900-square-foot house.)

Mensuration or biometrics (a) The measurement and calculation of volume, growth, and development of individual trees or stands and their timber products. (b) A measurement of forest lands.

Merchantable Logs exceeding a minimum size and a minimum usable volume that are suitable for sale.

Mixed stand   A timber stand in which less than 80 percent of the trees in the main canopy are of a single species.

Multiple use   The management of land or forest for more than one purpose, such as wood production, water quality, wildlife, recreation, aesthetics, or clean air.

Multiple-use forestry   Concept of forest management that combines two or more objectives, such as production of wood or wood-derivative products, forage and browse for domestic livestock, proper environmental conditions for wildlife, landscape effects, protection against floods and erosion, recreation, and protection of water supplies.

Multiple-use management   Management of land resources with the objective of achieving optimum yields of products and services from a given area without impairing the productive capacity of the site.

Natural regeneration   A stand of trees grown from natural seed fall or sprouting.

Natural resources   Land, water, and atmosphere, including their mineral, vegetable, and other components and flora and fauna on or in them.

Net annual growth   Increase in volume of trees during a specified year.

Net present value   The value of current investment discounted into the future, after all costs have been subtracted.

Noncommercial species   Tree species in which small size, poor form, or inferior quality is typical. These species do not normally develop into trees suitable for conventional forest products.

Nonindustrial private forest   Private forest land owned by an individual or organization that does not directly supply a production mill.

Nonpoint sources   Dispersed human activities such as agriculture, urban development, and forestry that cause water pollution (as opposed to point sources such as industrial effluent pipes).

Nontimber forest values   Values within the forest other than timber that include but are not limited to biological diversity, fisheries, wildlife, minerals, water quality and quantity, recreation and tourism, cultural and heritage values, and wilderness and aesthetic values.

Old-growth forest   Ecosystems distinguished by old trees and related structural attributes. Old growth encompasses the later stages of stand development, which typically differ from earlier stages in ways that may include tree size, accumulations of large dead woody material, number of canopy layers, species composition, and ecosystem function. The age at which old growth develops and the specific structural attributes that characterize old growth varies widely according to forest type, climate, site conditions, and disturbance regime.

Ownership   The identification of the legal owner/administrator (Federal, state, local, private) on both the surface and subsurface estates.

Partial cutting   Refers generically to harvest operations, under any of the several silvicultural systems, to cut selected trees and leave desirable trees for various stand objectives. Partial cutting includes harvest methods used for seed tree, shelterwood, selection, and clearcutting with patches of forest remaining.

Periodic annual increment   Mean annual growth or increase in volume during a specific period of time.

Planning   Determination of the goals and objectives of an enterprise and selection—through a systematic consideration of alternatives—of the policies, programs, and procedures for achieving them. An activity devoted to clearly identifying, defining, and determining courses of action before their initiation, necessary to achieve predetermined goals and objectives.

Plant or habitat diversity   A variety of food or cover for wildlife. Variation may occur at one point in time or over a period of time, such as during the course of a season. Seasonal diversity of food and cover is often critical to the survival of a species.

Precommercial thinning   Cutting trees from a young stand so that the remaining trees will have more room to grow to marketable size. Trees cut in a precommercial thinning have no commercial value, and normally none of the felled trees are removed for use. The primary intent is to improve growth potential for the trees left after thinning.

Prescribed or controlled burn   The use of fire under specific environmental conditions to achieve forest management objectives. Used to reduce hazardous fuel levels, control unwanted vegetation, favor desired vegetation, and improve visibility and wildlife habitat.

Preservation   The action of reserving, protecting, or safeguarding a portion of the natural environment from unnatural disturbance. It does not imply preserving an area in its current state, for natural events and natural ecological processes are expected to continue. Preservation is part of, and not opposed to, conservation.

Public involvement   The procedures for obtaining and considering the views of the general public in planning and decision-making processes.

Rangeland (range)   Land on which vegetation is predominantly grasses, forbs, or shrubs suitable for grazing or browsing. Rangeland is generally shrub land but may include some tree land and barren land. Agricultural land is excluded. Also included are areas seeded to native species or adapted introduced species that are managed like native vegetation.

Recreation opportunity class   A mix of outdoor settings based on remoteness, area size, and evidence of humans, which allows for a variety of recreation activities and experiences. The descriptions used to classify the settings are on a continuum and are described as rural, roaded resource, semiprimitive motorized, semiprimitive nonmotorized, and primitive.

Reforestation   Reestablishing a forest by planting or seeding an area from which forest vegetation has been removed.

Regeneration   The renewal of a tree crop through either natural means (seeded on site from adjacent stands or deposited by wind, birds, or animals) or artificial means (by planting seedlings or by direct seeding).

Regeneration cut   A cutting strategy in which old trees are removed while favorable environmental conditions are maintained for the establishment of a new stand of seedlings.

Reserve   An area of forest land that, by law or policy, is not available for harvesting. Areas of land and water set aside for ecosystem protection, outdoor and tourism values, preservation of rare species, gene pool, wildlife protection, and so on.

Residual stand   Trees left in a stand to grow until the next harvest. This term can refer to crop trees or cull trees.

Restoration   The return of an ecosystem or habitat to its original community structure, natural complement of species, and natural functions.

Riparian   An area of land adjacent to a stream, river, lake, or wetland that contains vegetation that, because of the presence of water, is distinctly different from the vegetation of adjacent upland areas.

Rotation   The number of years required to establish and grow trees to a specified size, product, or condition of maturity. For example, a pine rotation in the South may range from as short as 20 years for pulpwood to more than 60 years for sawtimber.

Roundwood   A length of cut tree generally having a round cross section, such as a log or bolt.

Sale unit   A timber sales arrangement in which the buyer pays for forest products removed in units (measured in cords, MBF, or units of weight).

Salvage cut   The harvesting of dead or damaged trees or of trees in danger of being killed by insects, disease, flooding, or other factors in order to save their economic value.

Sawlog or sawtimber   A log or tree that is large enough (usually 10 to 12 inches in diameter) to be sawed into lumber. Minimum log length is typically 8 feet.

Second growth   A forest or stand that has grown up naturally after removal of a previous stand by fire, harvesting, insect attack, or other cause.

Sedimentation   The deposition or settling of soil particles suspended in water.

Seed tree cut   A harvesting method in which a few scattered trees are left in the area to provide seed for a new forest stand. Selection of seed trees should be based on growth rate, form, seeding ability, wind firmness, and future marketability. This harvesting method produces an even-aged forest.

Selection system   Uneven-aged silvicultural system in which single or small groups of trees are periodically selected to be removed from a large area so that age and size classes of the reproduction are mixed.

Selective cutting   The periodic removal of individual trees or groups of trees to improve or regenerate a stand.

Shelterwood system   Even-aged silvicultural system in which a new stand is established under the protection of a partial canopy of trees. The mature stand is generally removed in a series of two or more cuts, the last of which is when the new even-aged stand is well developed.

Silviculture   The art and science of controlling the establishment, growth, composition, health, and quality of forests and woodlands. Silviculture entails the manipulation of forest and woodland vegetation in stands and on landscapes to meet the diverse needs and values of landowners and society on a sustainable basis.

Site class   The measure of the relative productive capacity of a site for a particular crop or stand, generally based on tree height at a given age and expressed as either good, medium, poor, or low.

Site preparation   Preparing an area of land for planting, direct seeding, or natural reproduction by burning or chemical vegetation control or by mechanical operations such as disking, bedding, scarifying, windrowing, or raking.

Slash   (a) Tree tops, branches, bark, or other residue left on the ground after logging or other forestry operations. (b) Tree debris left after a natural catastrophe.

Softwood (conifer)   A tree belonging to the order *Coniferales*. Softwood trees are usually evergreen, bear cones, and have needles or scalelike leaves. They include pine, spruces, firs, and cedars.

Soil compaction   A physical change in soil properties that results in a decrease in porosity and an increase in soil bulk density and soil strength.

Soil type   Soils that are alike in all characteristics, including texture of the topsoil. Soil maps and information on site index, erodibility, and other limiting properties are available from county Natural Resource Conservation Service offices.

Species   A group of related organisms having common characteristics capable of interbreeding.

Stand   An easily defined area of the forest that is relatively uniform in species composition or age and can be managed as a single unit.

Stewardship   Caring for land and associated resources and passing healthy ecosystems to future generations.

Stocking   A description of the number of trees, basal area, or volume per acre in a forest stand compared with a desired level for balanced health and growth. Most often used in comparative expressions—such as well-stocked, poorly stocked, or overstocked.

Stream flow   Measure of the volume of water passing a given point in a stream channel at a given point in time. Stream flow is a function of depth, width, and velocity of water in a channel. Changes in stream flow affect the available habitat for fish spawning or rearing.

Streamside management zone (SMZ)   An area adjacent to a stream in which vegetation is maintained or managed to protect water quality. The width depends on slope, but 50 feet is the normal minimum. Trees may be removed from SMZs as long as the streambed is not disrupted and sufficient vegetation is left to protect water quality.

Stumpage   The value or volume of a tree or group of trees as they stand uncut in the woods (on the stump).

Succession   The natural sequence of plant community replacement beginning with bare ground and resulting in a final, stable community in which a climax forest is reached. Foresters, wildlife biologists, and farmers constantly battle ecological succession to try to maintain a particular vegetative cover.

Sustainable forest management   Management regimes applied to forest land that maintain the productive and renewal capacities as well as the genetic, species, and ecological diversity of forest ecosystems.

Sustained yield   Management of forest land to produce a relatively constant amount of wood products, revenue, or wildlife.

Thinning   A tree removal practice that reduces tree density and competition between trees in a stand. Thinning concentrates growth on fewer higher-quality trees, provides periodic income, and generally enhances tree vigor. Heavy thinning can reduce fire risk by reducing the amount and location of hazardous fuels. Thinning can also benefit certain types of wildlife by increasing the growth of ground vegetation.

Threatened or endangered species   A threatened species exhibits declining or dangerously low populations but still has enough members to maintain or increase numbers. A species is endangered when the total number of remaining members may not be sufficient to reproduce enough offspring to ensure survival of the species.

Timber   General term applied to forests and their products.

Tree farm   A privately owned forest or woodland in which timber crop production is a major management goal. Many tree farms are officially recognized by the American Tree Farm System, an organization sponsored by the American Forestry Council.

Understory   (a) The layer formed by the crowns of smaller trees in a forest. (b) The trees or foliage layer beneath the forest canopy.

Uneven-aged forest   Theoretically, these stands contain trees of every age on a continuum from seedlings to mature canopy trees. In practice, uneven-aged stands are characterized by a broken or uneven canopy layer. Usually the most trees are in the smaller-diameter classes. As trees increase in diameter, their numbers diminish throughout the stand. See also *Even-aged forest.*

Uneven-aged management   Silvicultural system in which individual trees originate at different times, resulting in a forest with trees of all ages and sizes. Trees are harvested on an individual basis.

Urban-wildland interface   Areas where urban development encroaches on undeveloped lands; specifically, an area of heightened concern over wildland forest fire management and impacts.

Water quality   The physical, chemical, and biological properties of water.

Watershed   An area of land that collects and discharges water into a single main stream through a series of smaller tributaries.

Wetland   Areas inundated by surface water or groundwater frequently enough to support—and under normal circumstances they do or would support—a prevalance of vegetation or aquatic life that requires saturated soil conditions for growth and reproduction. Wetlands generally include swamps, marshes, bogs, and similar areas such as sloughs, potholes, wet meadows, river overflows, mudflats, and natural ponds.

Wilderness   An area of land intentionally set aside and designated the highest protections from human activity and development.

Wildfire   Unplanned fire requiring suppression. Can be contrasted with a prescribed fire that burns within prepared lines, enclosing a designated area, under predetermined conditions.

Wildlife   A broad term that includes nondomesticated vertebrates, especially mammals, birds, and fish.

Withdrawals   Lands that have been removed or segregated from the operation of some or all of the public land laws through executive or Congressional action.

Index

Accredited forestry schools, 220–226

Acre (defined), 277

Adaptive management (defined), 277

Adirondack State Park, 75, 154

Administrative procedures, 67

Adopt-A-Stream Foundation, 197

Afforestation (defined), 277

Age class (defined), 277

Agenda 21 (UNCED), 85

Alabama Forestry Commission, 190

Alaska
   comprehensive private forestry regulation, 66(table)
   forest ecoregion, 144
   forest practices programs, 70
   Wilderness areas, 153

Alaska Division of Forestry, 190

Alaska National Interest Lands Conservation Act (1980), 15, 42, 43, 126, 140

Alaska Purchase (1867), 6, 117

Aldo Leopold Foundation Inc., 197

All-aged or uneven-aged management (defined), 277

All-aged or uneven-aged stand (defined), 277

Allowable cut (defined), 277

Amazon, 128

Ambio, 98

Ambler Realty Company, Village of Euclid v., 64

American Association for the Advancement of Science, 8, 134

American chestnut, 105

American Conservation Association Inc., 198

American elm, 105

American Fisheries Society, 232

American Forest and Paper Association (AF&PA), 229–230
   Expert Review Panel, 86
   Sustainable Forestry Board, 86
   Sustainable Forestry Initiative (SFI), 32, 86, 87, 175–176

American Forest Congress (1905), 120

American Forest Congress (1995), 128

American Forest Foundation (AFF), 176, 234

American Forest Resource Council (AFRC), 234

American Forestry Association, 8

American Forests, 55, 198

American Hiking Society, 198

American Lands, 198–199

American Revolution, 5

American Samoa Forestry Program, 190

293

American Tree Farm System (ATFS), 234
  Standards, Guidelines, and Performance Measures for certification, 176, 177–179(table)
  web site, 176
Anadromous fish (defined), 277
Ancient Forest International, 199
Animal and Plant Health Inspection Service (APHIS), 103–104
Animal unit month (AUM) (defined), 277
Annual allowable harvest (defined), 278
Annual growth (defined), 278
Applegate Partnership, 55
Applied forestry, 261
Aquatic habitat (defined), 278
Area regulation (defined), 278
Arizona, 47
Arizona State Land Department, 190
Arkansas Forestry Commission, 191
Artificial regeneration, 278
Asian long-horned beetle, 161
Association of Consulting Foresters of America Inc., 232
Auburn University, 221
Australia, 99

Backfire (defined), 278
Bailey, Robert G., 144
Ballinger, Richard A., 121
Ballot initiatives, 74
Bandelier National Monument, fire in, 50
Bari, Judi, 272
Basal area (defined), 278
Belize: Rio Bravo Conservation and Management Area, 97–98
Benefit/cost analysis (defined), 278

Best management practices (BMP), 71
  defined, 278
  nonprint resources, 274
Bill of Rights, 4
Bingaman, Jeff, 59
Biodiversity
  books and reports, 241, 242, 251
  defined, 278
  indicators of, 162–163
  nonprint resources, 274
  online reports, maps, and information, 267
  performance measures, 178(table)
  standards for, 178(table)
Biodiversity conservation, 33–41, 107–108, 162–163
  in forest plantations/protected areas, 38–40
  global strategy for, 109
  Montreal Process criterion, 162–163
  in private lands, 37–38
  in public lands, 35–37
  recent policy developments, 35–40
  scientific journals, 252–253
Biological diversity. *See* Biodiversity
Biology
  conservation, 279
  scientific journals, 254
Biomass (defined), 278
Biomass Users Network, 199
Biometrics (defined), 285
Bioremediation, 104
Biotechnology
  books and reports, 251
  forest, 101–105
  microbial products of, 104
  recent policy developments, 103–104
  scientific journals, 253
Birch, Thomas W., 156
Bitterroot National Forest, 131, 136

Board foot (defined), 278
Bob Marshall Wilderness Area,
    136
  fire in, 50
Boise Cascade, 227
Bolle, Arnold, 131
"Bolle Report," 131
Books and reports
  general forestry and forest
    conservation, 240–242
  international forestry and
    forest conservation, 250–252
  U.S. forest history, 248–250
  U.S. forest policy, 243–247
Bosworth, Dale, 47
Boundary Waters Canoe Area,
    123
Bowater Inc., 227
Brazil: Carbon Sink Project
    (Matto Grosso), 98
Broad Arrow policy, 5, 117
Broadcast burning (defined), 278
Broadleaf (defined), 278
"Brundtland Report," 252
Buffalo Creek fire (Colorado), 93
Buffer strip (defined), 279
Bureau of Indian Affairs (BIA),
    88, 146–147, 157, 188
Bureau of Land Management
    (BLM), 151–153, 188
  area of forest land and timber
    land owned, 147(table)
  Church Subcommittee
    Guidelines, 132
  conserving biological
    diversity, 36
  forest lands, 147
  management of Wilderness
    areas, 153
  ownership of forests, 146
  precursor, 135
  state offices, 151–153
  stewardship contracts, 52
  web site, 151
Burning
  broadcast, 278
  intentional, 4

let-burn policies, 50
  prescribed or controlled, 50,
    279, 287
  See also under Fire danger; Fire
    frequency; Fire suppression
Bush, George H. W., 99
Bush, George W., 52
  and Clinton Administration's
    Roadless Rule, 44
  Healthy Forests Initiative, 52
  withdrawal from Kyoto
    Protocol, 99
Butz, Izaak Walton League of West
    Virginia v., 15–16, 65, 134
Butz, Sierra Club v., 43

Calhoun, John, 118
California
  early forest policy, 7
  forest practices act, 69–70
  forest practices programs, 70
  large tracts of low-access
    forest, 160
  private forestry regulation,
    63(table), 66(table), 74
  purchase of Federal
    timberlands, 119
California Department of
    Forestry and Fire Protection,
    191
California Polytechnic State
    University, 221
Campbell Group, 155
Can Tropical Forests Be Saved?
    (video), 270
Canada
  carbon credit trading system,
    99
  trade disputes, 121
Canada–United States
    Environmental Council, 199
Canadian Parks and Wilderness
    Society, 200
Canadian Standards Association,
    85
Canadian Wildlife Federation,
    200

Canadian Woodlands Forum, 253
Canyon Creek fire (Montana), 50
Cap-and-trade system, 99
Capper, Arthur, 122
Capper Report, 122
Carbon
  atmospheric, 94–101
  cap-and-trade system, 99
  conservation, 95
  credit trading system, 99, 106, 129
  forest conservation activities that help reduce, 95
  global cycles, 166
  reduction projects, 97–98
  recent policy developments, 97–100
Carbon dioxide (CO_2) emissions, 99, 128–129
Carbon sequestration, 95
Carbon Sink Project (Matto Grosso, Brazil), 98
Carbon sinks, 98, 129
Carbon substitution, 95
Carrying capacity (defined), 279
Carson, Rachel: *Silent Spring*, 2
Categorical Exclusion (CE) projects, 52
Center for Watershed Protection, 200
Central Pacific Railroad, 7, 118
Central Park (New York City), 137
Certification
  for chain of custody, 172–173, 174
  defined, 83
  forest, 171–179, 282
  forest management, 82–89, 143
  green, 39, 84
  Tree Farm, 177–179(table)
  wood, 171–173, 174
Certified Forest Products Council, 235
Chain of custody, certification for, 172–173, 174

Chemicals, prudent use of
  performance measures, 178(table)
  standards for, 178(table)
Chesapeake Bay Critical Areas Protection Act (1984), 72
Chicago Climate Exchange (CCX), 99
Chile: Reserva Forestal Malleco, 98
Chip (defined), 279
Chugach National Forest, 43, 121
Church, Frank, 131–132
Church Subcommittee Guidelines, 131–132
Clarke-McNary Act (1924), 123, 134
Class I areas, 124
Clean Air Act (1955), 16, 124
Clean Air Act Amendments of 1970, 2–3
Clean Air Act Amendments of 1990, 99
Clean Water Act (1972), 2, 16, 70, 91
Clean Water Act Amendments of 1977, 2
Clean Water Act Amendments of 1987, 91
Clearcut harvest (defined), 279
Clearcutting
  ballot initiatives to ban, 74
  controversy, 15–16, 65, 131
  Oregon regulatory standards, 69
Clemson University, 222
Cleveland, Grover, 139
Clinton, Bill, 44
Clinton Administration
  Roadless Area Rule, 44, 160
Colorado Division of Wildlife Habitat Partnership Program, 60
Coastal Alliance, 200–201
Coastal Society, 201
Collins Companies, 227

Colorado
    forest health, 47
    forestry and water, 93
    recent policy developments, 93
    state forest lands, 154
    See also Hayman fire case
        study analysis
Colorado River watershed, 138
Colorado State Forest Service,
    191
Colorado State University, 222
"Combating Deforestation"
    (UNCED), 85
Commercial thinning (defined),
    279
Common law, English, 4
Community-Based Forest and
    Public Lands Restoration
    Act (did not pass), 59
Community forestry, 53–61
    basic elements of, 56–58
    books and reports, 245–246
    current policy developments,
        58–61
    defined, 279
    global movement toward,
        57–58
    Scolel Te Community Forestry
        Project, Mexico, 98
Composite Panel Association,
    230
Conifer forests, 144, 279(defined)
Connecticut, 5
Connecticut Division of Forestry,
    191
Connecticut Forest Practices Act
    (1991), 72–73
Conservation
    biodiversity, 33–41, 107–108,
        162–163
    books and reports, 240–242,
        249
    carbon, 95
    of corporate forest lands,
        77–79
    defined, 279
    of family forest lands, 79–82

forest, 95, 168–170, 240–242,
    250–252, 266
forest land, 74–82
indicators of, 165
Pinchot as martyr for, 121
of public forest lands, 75–77
Rio Bravo Conservation and
    Management Area, Belize,
    97–98
scientific journals, 252–253,
    254, 259, 264
of soil and water resources,
    165
Conservation biology
    defined, 279
    scientific journals, 254
Conservation bonds, 79
Conservation easements, 81–82
    defined, 279
    "working forest," 79
Conservation ethic, 8–9
Conservation Fund, 78
Conservation International, 201
Conservation Law Foundation,
    201
Conservation Movement, 1–22
    leaders, 135–136, 138, 139
Conservation organizations,
    nongovernment, 197–220
Conservation Reserve Program,
    40
Constitution. *See* U.S.
    Constitution
Continental Congress, 5–6
Contingent regulation, 71
Controlled burning, 50,
    279(defined), 287
Convention on Biological
    Diversity (1992), 3
*Conversations on Sustainable
    Forestry* (video), 273
Coolidge, Calvin, 123
Cooperative Extension Service,
    61
Cooperative State Research,
    Education, and Extension
    Service, 186

Cooperative sustained-yield
    units, 124
Cord (defined), 279
Cornell School of Forestry,
    15–16, 133
Corporación Nacional Forestal
    (Chile), 98
Corporate forest lands, 77–79
Cost analysis, 278
Cost-share assistance (defined),
    280
Council on Environmental
    Quality, 184
County forest lands, 154
County Supremacy Movement,
    76
Coverts program, 61
Craig, Larry, 59
Creative Act (1891), 148
Criteria and Indicators (C&I)
    Montreal Process, 128, 143,
        162–168, 168–170
    Working Group, 161
Critical habitat (defined), 280
Cruise (defined), 280
"Culmination of mean annual
    increment" rule, 26
Cultural, social, and spiritual
    needs and values, 167
Cumulative effects (defined), 280
Custodial management, 11–12,
    27–28
Cut, allowable, 277
Cut-and-run practice, 17, 28, 54
Cut-out-and-get-out practices,
    122
Cutting cycle (defined), 280
Cutting patterns, 90
Cuyahoga River, 66

Dana, Samuel Trask, 132
Death taxes, 19, 81
Deciduous forests, 144
Deciduous trees, 284
Decision-making
    decentralized, 59–60
    participatory processes, 57

Declaration of Independence, 5
Defenders of Wildlife, 201–202
Deforestation
    "Combating Deforestation"
        (UNCED), 85
    early, 7
    tropical, 127, 128
Delaware Forest Service, 191
Dendrology, 280
Development. *See* Research and
    development
*Dexter, Avery, State of Washington
    v.*, 64–65
Diameter at breast height (DBH)
    (defined), 280
Diameter classes (defined), 280
Diameter limit (defined), 280
Dingley Tariff Act (1897), 121
District of Columbia Trees and
    Lands, 191
Diversity
    biological. *See* Biodiversity
    genetic, 163, 283
    plant or habitat, 287
    species, 163
Dogwood Anthracnose, 105
Douglas fir, 48
Drinking water quality, 96
Dry domain, 144
Duke University, 222
Duluth Timber Company, 228
Dutch Elm Disease, 105
Dwyer, William, 132

*Earth as Modified by Human
    Action* (Marsh), 118–119
Earth Report (video), 270
Earth Summit. *See* United
    Nations Conference on
    Environment and
    Development
Earthstewards Network, 202
Earthwatch Institute, 202
Eastern Shore Land
    Conservancy, 202–203
Eastern Wilderness Act (1974),
    43, 126

Ecolabels, 83
Ecological health (defined), 280
Ecological integrity (defined), 280
Ecological type (defined), 280–281
Ecology
  defined, 281
  landscape, 285
  scientific journals, 258, 263
Economic Action Program, 51, 58
Economic framework, 168–170
  books and reports, 241, 246
  indicators of, 169–170
Ecoregions
  books and reports, 242
  forest, 144–145
  online reports, maps, and information, 268
Ecosystem(s)
  defined, 281
  indicators of diversity, 162–163
  indicators of health and vitality, 164–165
  indicators of productive capacity, 163–164
  maintenance of health and vitality of, 164–165
  maintenance of productive capacity of, 163–164
  scientific journals, 254
  sustainability of, 24
Ecosystem management
  books and reports, 242, 245–246
  nonprint resources, 274, 275
Edge (defined), 281
El Colegio de la Frontera, 98
Employment needs, 167–168
Endangered or threatened species, 127, 281(defined), 290
Endangered Species Act (1973) (ESA), 16, 35, 65, 125, 148
  as barrier to timely action, 52
  requirements for private lands, 37

requirements for public lands, 35–37
Enforcement, 68
England
  Broad Arrow policy, 117
  common law, 4
  See also United Kingdom
Environment (defined), 281
Environmental awareness, era of, 1–3
Environmental Defense, 203
Environmental impact analysis
  books and reports, 241
  Categorical Exclusion (CE) projects, 52
  public concerns over, 72–73
Environmental impact statement (EIS), statewide, 73–74
Environmental Law and Policy Center of the Midwest, 203
Environmental law journals, 253–254, 254–255, 256–257, 258–259, 260, 262–263
Environmental management journals, 255, 258
Environmental News Service, 265
Environmental organizations, 266
Environmental Protection Agency. *See* U.S. Environmental Protection Agency
Environmental Quality Improvement Program, 60
Erosion (defined), 281
Estate taxes, 19, 81
Euclid, Village of, v. Ambler Realty Company, 64
Europarc Federation, 203–204
Europe, 25–26
European Union, 99
Even-aged forest (defined), 281
Even-aged management (defined), 281
Exotic pests, 161
Exotic species, 161
Extinction rates, 33

Family forest lands, 79–82
Farm Bill (1996), 60
Farm Bill (2002), 60
Farm use, 158(table)
Fate of the Forest (video),
   270–271
Federal Forest Reserves, 54–55,
   120
   first 13, 139
   increase of, 139
   purpose of, 27
   renamed National Forests, 120
   See also National Forest System
Federal public forest lands,
   147–153
   books and reports, 244, 248
   See also National Forest System
Federal Trade Commission
   (FTC), 84
Federal Water Pollution Control
   Act (1948), 91, 124
Federal Water Pollution Control
   Act (1972), 66–67
Fernow, Bernhard, 15–16, 27,
   132–133
Fire(s). *See under location of
   specific fire*
Fire danger
   defined, 281
   information clearinghouses,
   266, 267
   See also Wildfire(s)
Fire frequency, 160–161
Fire hazard (defined), 281
Fire suppression, 48, 148
   defined, 281
   federal costs, 47
   National Fire Plan, 51, 56, 58
   by Native Americans, 48
   policy of, 11
   Ten A.M. Rule, 48
   videos, 272
Firebreak (defined), 281
Firefighting, 11, 48
"First in time, first in right"
   allocations, 91

Fish, anadromous (defined), 277
Florida
   concerns over water quality, 71
   early private forestry
   regulation, 63(table)
   forestry and water, 93
   recent policy developments, 93
Florida Division of Forestry, 191
Florida Everglades, 144
Florida Purchase (1819), 6, 117
Food and Agriculture
   Organization (FAO), 268, 269
Forage (defined), 282
Ford Foundation, 57
Forest(s)
   and atmospheric carbon,
   94–101
   as carbon sinks, 129
   conditions and trends, 158–161
   defined, 282
   exploitation of, 6–7
   facts and data, 143–181
   indicators of consumption,
   166–167
   long-term socioieconomic
   benefits of, 166–167
   performance measures,
   178(table)
   standards for aesthetics of,
   178(table)
   U.S. ownership, 146–147
   and water, 89–93
   See also National Forest System
Forest and Rangelands
   Renewable Resources
   Planning Act (1974) (RPA),
   29, 126, 134
Forest area, 145–146
Forest biotechnology, 101–105
Forest certification
   defined, 282
   principles and standards,
   171–179
Forest conservation
   activities that help reduce
   atmospheric carbon, 95

books and reports, 240–242,
250–252
information clearinghouses,
266
international, 250–252
legal, institutional, and
economic framework for,
168–170
Forest Conservation Act
(Washington), 64
Forest Conservation Portal, 266
Forest cover type (defined), 282
Forest-cutting patterns, 90
Forest ecoregions, 144–145
Forest Ecosystem Management
Assessment Team (FEMAT),
140
Forest ecosystems
indicators of health and
vitality, 164–165
indicators of productive
capacity, 163–164
maintenance of health and
vitality of, 164–165
maintenance of productive
capacity of, 163–164
sustainability of, 24
Forest Fragmentation (video), 273
Forest health, 47–53
defined, 282
recent policy developments,
50–52
Forest History Society, 233
Forest industry, 154
area of forest land and timber
land owned, 147(table)
books and reports, 250
investment in, 167
scientific journals, 263
Forest Investment Associates,
155
Forest land
county, 154
defined, 282
federal acquisition of, 121
large tracts of, 159–160

online reports, maps, and
information, 268
standards and guidelines for
owners, 177–179(table)
state, 153–154
U.S. ownership, 146–147,
147(table)
Forest land area, 145
Forest land conservation, 74–82
Forest Legacy program, 78
Forest management
in biodiversity conservation,
107–108
books and reports, 241
custodial, 27–28
defined, 282
early laws, 5–6
information clearinghouses,
267
intensive, 284
international initiatives, 143
policies in National Forests,
11–16
on private lands, 18–20
proposals for action, 129
scientific journals, 255
sustainable, 16–20, 24–32,
168–170, 290
sustained-yield, 122
Forest management certification,
82–89, 143
objectives, 84
recent policy developments,
86–89
Forest management plan
(defined), 282
Forest planning, 135
Forest plantations and protected
areas
books and reports, 250
consensus agreement on, 40
conserving biological diversity
in, 38–40
early, 119
expansion, 109
trends in, 160

Forest policy
  books and reports, 243–247
  current issues in, 23–116
  introduction to, 4–5
  in National Forests, 11–16
  roots of, 1–22
  scientific journals, 255
  U.S., 23–116
Forest practice (defined), 282
Forest practices acts (FPAs), 73
  basic components, 67–68
  state-level, 67–68
"Forest Principles" (UNCED),
  31–32, 85
Forest production, indicators of,
  166–167
Forest products, 145–146
  globalization of, 58
  scientific journals, 256
Forest products companies,
  227–229
Forest products industry, 77–78
  evolution of, 17–18
  international consolidation of,
  127
Forest Products Society, 230
Forest Reserve Act (1891), 8–9,
  27, 75, 120, 133, 134
Forest resource councils (FRCs),
  74
Forest resources
  defined, 282
  online reports, maps, and
  information, 269
Forest Resources Association
  Inc., 230
Forest Service Employees for
  Environmental Ethics, 204
Forest Stewards Guild, 235
Forest stewardship, 56–57
*Forest Stewardship at the Urban/
  Rural Interface* (video), 275
Forest Stewardship Council, U.S.
  (FSC), 39, 87, 171, 235
  formed, 127
  Principles and Criteria, 85
  purpose, 85

Forest stewardship plan
  (defined), 282
Forest Trust, 55, 204
Forest types, 144–145, 282–283
Forest water quality, 91
ForestInformation.com, 266
Forestland Group, 155
Forestry
  accredited schools, 220–226
  books and reports, 240–242,
  248–250, 250–252
  community, 279
  defined, 283
  international, 250–252
  multiple-use, 12–14, 28–29, 286
  as point source of water
  pollution, 92
  private regulation, 62–74
  professional associations,
  232–233
  scientific journals, 252–264
  sustainable, 4–5, 24
  sustained-yield, 27
  trade associations, 229–232
  U.S., 248–250
  videos, 270–272
  world, 251
Forestry Commission, 139
Forestry contractors
  performance measures,
  179(table)
  standards for use, 179(table)
Forestry Media Center (Oregon
  State University College of
  Forestry), 273–274
Forestry-related associations,
  234–237
*Forests and Water in the Light of
  Scientific Investigation* (Zon),
  141
Forests.org, Inc., 266
Fragmentation
  defined, 283
  nonprint resources, 273
Franklin, Jerry, 133
Fraud and abuse, 119, 120
Free Timber Act (1878), 7, 119

French Fund for the
Environment, 98
Friends of the Earth, 204
Fuel loading (defined), 283
Fuel management (defined), 283
Fuelbreak (defined), 283
Fuels treatment projects, 52

Gadsden Purchase (1853), 6, 117
"Gang of Four," 133, 135, 140
Gene mapping, 103
General Agreement on Tariffs
and Trade (GATT), 85
General Land Ordinance (1785), 6
Genetic diversity
defined, 283
indicators of, 163
Genetic engineering, 102–103
Genetically modified (GM) trees,
103, 104
Geographic information system
(defined), 283
Georgia
local ordinances, 71
western reserve claim, 5
Georgia Forestry Commission,
191
Georgia Pacific, 78, 228
Germany: trade restrictions, 85
Global Association of Online
Foresters (GAOF), 236
Global carbon cycles
indicators of forest
contribution to, 166
maintenance of forest
contribution to, 166
Global resource management
books and reports, 250
online reports, maps, and
information, 268
Globalization of forest products,
58
Gordon, John, 133
Grade (defined), 283
Grand Canyon Trust, 205
Grassland(s)
books and reports, 244–245

defined, 283
National Grasslands, 148
Grassland Heritage Foundation,
205
Graves, Henry, 133–134
Grazing, 134–135, 137
Grazing Service, 135
Greater Yellowstone Coalition,
205
Green certification, 39, 84
Green strip (defined), 283
Greenhouse gases, 1–2
Greenpeace, 272
Group selection (defined), 283
Growing stock (defined), 283
Guam Forestry and Soil
Resources Division, 192
Gypsy moth, 161

Habitat
aquatic, 278
critical, 280
defined, 284
performance measures,
177(table)
standards for, 177(table)
Habitat diversity (defined), 287
Habitat Partnership Program, 60
Hanzlik's Formula, 26
Hardwood Plywood and Veneer
Association, 231
Hardwoods
defined, 284
trends in, 160
Harvest schedule (defined), 284
Hawaii, 144
Hawaii Division of Forestry and
Wildlife, 192
Hayman (Colorado) fire case
study analysis, 269
Headwaters Forest, 272
Healthy Forests Initiative, 52,
129
Heartwood, 205–206
Help for Your Woodlands (video),
274
High-grading (defined), 284

Hill, James J., 140
Hill, Matthew W., 65
Historical range of variability
    (defined), 284
History, U.S. forest, books and
    reports, 248–250
Home Depot, 228
Homestead Act (1862), 6, 118
Hoover Commission, 132
Hot spots, 35
Hough, Franklin, 134
Hubbard Brook, 90
*Huckleberry Story, The: Building a
    Bridge between Culture and
    Science* (video), 273
Human dimension (defined),
    284
Humboldt State University, 222
Humid Temperate domain, 144
Humid Tropical domain, 144
Humphrey, Hubert, 134
Hydrology (defined), 284

Ickes, Harold, 134–135
Idaho
    forest health, 47
    forest practices programs, 70
    large tracts of low-access
        forest, 160
    private forestry regulation,
        63(table), 66 (table)
    roadless areas in, 43, 44, 46
    Wilderness protection, 132
Idaho Department of Lands, 192
Illinois: local ordinances, 71
Illinois Division of Forest
    Resources, 192
Import bans on tropical timbers,
    127
Import tariffs and duties, 18, 121
Improvement cut (defined), 284
Incentive (defined), 284
Incidental taking, 37
Indian Forest Management
    Assessment (2002), 88
Indiana Department of Natural
    Resources, 192

Indicator species (defined), 284
Industrial private forest
    (defined), 284
Information clearinghouses,
    265–267
Information online, 267–270
Institutional framework,
    168–170
Instituto pro Natura (Brazil), 98
Instream flow rights, 91
Intensive forest management
    (defined), 284
International Ecology Society,
    206
International forestry and forest
    conservation, 250–252
International Paper, 78, 99
International Society for the
    Preservation of the Tropical
    Rainforests, 206
International Standards
    Organization (ISO), 85–86
International Union for
    Conservation of Nature and
    Natural Resources, 206
International Wood Products
    Association, 231
Internet
    information clearinghouses,
        265–267
    online reports, maps, and
        information, 267–270
Inventory and Monitoring
    Institute, 266–267
Investment
    in forest sector, 167
    land, 157, 158(table)
Iowa Department of Natural
    Resources, 192
Iowa State University, 222
Island Resources Foundation,
    206–207
Izaak Walton League of America
    Inc., 207
*Izaak Walton League of West
    Virginia v. Butz*, 15–16, 65,
    134

John Hancock Timber Resource
  Group, 155
Johnson, K. Norman, 135
Journals, scientific, 252–264

Kansas, Mugler v. State of, 64
Kansas Forest Service, 192
Kentucky Division of Forestry,
  192
Keystone species (defined),
  284–285
Knutson-Vandenberg Act (1930),
  123
Kyoto Protocol, 94, 97, 128–129
  online report, 270
  U.S. withdrawal from, 99, 129

L-20 regulations, 42
Ladder fuels, 49, 285(defined)
Land and resource management
  plan (defined), 285
Land and resource planning, 247
Land and Water Conservation
  Fund (LWCF), 76
Land and Water Conservation
  Fund Act (1965), 125
Land conservation, 96
Land ethics, 135
Land grants, 118, 119
Land investment, 157, 158(table)
Land stewardship contracts,
  58–59
Land trusts, 78–79
Landscape (defined), 285
Landscape ecology
  defined, 285
  scientific journals, 259–260
Large tracts, of forests, 159–160
Law journals, 254
Leave strip (defined), 285
Lectures in Forest Policy
  (Schenck), 139
Legal framework, 168–170
Leopold, Aldo, 15, 42, 135
Let-burn policies, 50
Lieberman, Joseph, 100
Local ordinances, 71–72

Log rule (defined), 285
Log scale (defined), 285
*Logging, Best Management
  Practices and Water Quality*
  (video), 274
Logging journals, 253
Long-term benefits, indicators
  of, 166–167
Louisiana, 63
Louisiana Office of Forestry, 193
Louisiana Purchase (1803), 6, 117
Louisiana State University, 222
Low-access forest land, 159–160
Lumber trade disputes, 121

Magna Carta, 4
Maine: private forestry
  regulation, 74
Maine Forest Practices (1989), 72
Maine Forest Service, 193
Man and Nature (Marsh), 8,
  118–119, 135
Management
  adaptive, 277
  all-aged or uneven-aged, 277
  best management practices
    (BMP), 71, 274, 278
  custodial, 11–12, 27–28
  ecosystem, 242, 245–246, 274, 275
  even-aged, 281
  forest, 107–108, 129, 241, 255,
    267, 282, 284
  fuel, 283
  global resources, 250, 268
  intensive, 284
  international initiatives, 143
  multiple-use, 243, 286
  of National Forests, 11–16,
    27–28, 138, 248–249
  on private lands, 18–20
  resources, 273–274
  sustainable, 24–32, 35, 85–86,
    98, 168–170, 177(table),
    244–245, 290
  sustained-yield, 122
  uneven-aged, 277, 291
  woodland, 273–274

Management plan/planning
defined, 285
nonprint resources, 273–274
*Management Planning for Your
Small Woodland: An
Introduction* (video), 273–274
*Managing for Biodiversity in
Young Forests* (video), 274
*Managing Forest Ecosystems:
Assessing New Opportunities*
(video), 275
Maps, online, 267–270
Marginal land (defined), 285
Marsh, George Perkins
biographical sketch, 135–136
*Man and Nature (Earth as
Modified by Human Action)*, 8,
118–119, 135
Marshall, Robert, 15, 42, 136
Maryland
Chesapeake Bay Critical Areas
Protection Act (1984), 72
concerns over water quality,
71
early private forestry
regulation, 63(table)
local ordinances, 71
private forestry practices, 72
Maryland Cooperative Extension
Service, video by, 274
Maryland Department of
Natural Resources–Forest
Service, 193
Massachusetts
forest legislation, 117
forest practices programs, 70
private forestry regulation,
63(table), 66(table)
Massachusetts Department of
Environmental
Management, 193
Massachusetts Forest Cutting
Practices Act (1983), 66
Mather, Stephen T., 42, 122, 136
Mature tree (defined), 285
MBF (defined), 285

McCain, John, 100
McGuire, John, 136
McKinley Tariff Act (1890), 121
Mead Westvaco Corporation, 99,
228
Measurement, capacity for, 170
Menominee Tribal Enterprises,
176
Mensuration (defined), 285
Merchantable (defined), 285
Metcalf, Lee, 131
Mexican Cession (1848), 6, 117
Mexico: carbon credit trading
system, 99
Michigan
forest ecoregion, 144–145
timber trespass, 7
total forest land area, 159
Michigan Department of
Natural Resources Forest
Management Division, 193
Michigan State University, 222
Michigan Tech University, 222
Mid-Atlantic Council of
Watershed Associations, 207
Mineral exploration, 119
Mining Law (1872), 119
Ministry of Water, Lands, and
Air Protection (Canada),
184–185
Minnesota
environmental impact
statement (EIS), 73–74
forest ecoregion, 144–145
large tracts of low-access
forest, 160
total forest land area, 159
Minnesota Division of Forestry,
193
Minnesota Forest Resource
Council (FRC), 74
Mississippi: early private
forestry regulation, 63(table)
Mississippi Forestry
Commission, 193
Mississippi State University, 222

Missouri: early private forestry
    regulation, 63(table)
Missouri Department of
    Conservation, 193
Mixed stand (defined), 286
Monitoring, 57
    capacity for, 170
    information clearinghouses,
        266–267
    multiparty, 60
Monongahela National Forest,
    15–16, 134, 136
Montana
    concerns over water quality, 71
    forest health in, 47
    large tracts of low-access
        forest, 160
    roadless areas in, 43
Montana DNRC–Forestry
    Division, 194
Montreal Process, 32, 110,
    161–170
    Criteria and Indicators (C&I),
        128, 143, 162–168, 168–170
Moore, Patrick, 272
Mugler v. State of Kansas, 64
Muir, John, 136–137
Multiparty monitoring, 60
Multiple use (defined), 124, 286
Multiple-use forestry, 12–14,
    28–29, 286(defined)
Multiple-use management
    books and reports, 243
    defined, 286
Multiple-Use Sustained-Yield
    Act (1960), 13–14, 28–29, 124,
    148
Municipal forest lands, 154
Murie, Olaus, 137

National Academy of Sciences
    Forestry Commission, 139
National Arbor Day Foundation,
    207
National Association of
    Professional Forestry

Schools and Colleges
    (NAPFSC), 236
National Association of
    Recreation Resource
    Planners, 207–208
National Association of State
    Foresters, 190
National Audubon Society, 208
National Commission on Science
    for Sustainable Forestry, 267
National Conference on Outdoor
    Recreation, 123
National Conservation
    Foundation, 208
National Environmental Policy
    Act (1970), 65, 125, 148
National Environmental Policy
    Act (1970) (NEPA), 2, 16, 52
National Fire Plan, 51, 56
    Economic Action Program, 51,
    58
    10-Year Implementation Plan,
    51, 56
National Forest Foundation, 208
National Forest Management
    Act (1976) (NFMA), 16, 30,
    65–66, 126, 131, 148
    as barrier to timely action, 52
    basis for, 132
    habitat conservation
        requirements, 36–37
    implementing, 137
    lead sponsor, 134
National Forest System (NFS), 9,
    10(figure), 28, 75, 120, 147,
    148–150
    area of forest land and timber
        land owned, 147(table)
    books and reports, 244–245,
        246, 247, 248–249, 249–250
    clearcutting controversy, 15–16
    conserving biological diversity
        in, 36–37
    created, 9
    custodial management of,
        11–12, 27–28

National Forest System (*continued*)
  expansion of, 75
  forest health, 47
  forest management policies,
    11–16
  intensive use of, 11–12
  low-access forest land, 159–160
  management of, 138, 248–249
  multiple use of, 12–14,
    124–125, 147
  plan revisions, 44–45
  purpose, 27
  recreation use of, 12–13,
    14(figure), 122
  reforestation of, 123
  regional offices, 148–150
  risk from catastrophic
    wildfires, 159
  silvicultural practices, 123
  timber harvest, 12
  timber harvest from, 13(figure)
  timber salvage sales, 128
  water rights, 91–92
  web site, 148
  Wilderness areas, 14–15, 43
National Forestry Association,
  209
National Forestry Program
  Committee, 122–123
National Grasslands, 148
National Hardwood Lumber
  Association, 231
National Indian Forest Resource
  Management Act (NIFRMA)
  (1990), 88
National Interagency Fire
  Center, 267
National Monuments, 126
National Network of Forest
  Practitioners (NNFP), 55,
  209, 236
National Oceanic and
  Atmospheric Administration
  (NOAA), 187
National Outdoor Recreation
  Review Commission, 132

National Park and Conservation
  Association, 209
National Park Service, 13, 188
  controlled burns, 50
  established, 122
  founder, 136
  management of Wilderness
    areas, 153
  ownership of forests, 146
National Park System (NPS), 13,
  123, 147, 153
  forest lands, 148
  recreation visits, 12–13,
    14(figure)
  scientific journals, 260
  web site, 153
  See also specific parks
National Park Trust, 209–210
National Science Foundation,
  185
National Tree Trust, 237
National Wilderness
  Preservation System
  (NWPS), 15, 147, 148, 153
  created, 125, 137
  web site, 153
National Wildlife Federation,
  210
National Woodland Owners
  Association (NWOA), 237
Native Americans
  forest management
    certification, 87–88
  forest policy, 4, 48
Native Ecosystems Council, 210
Native Forest Network, 210
Natural regeneration (defined),
  286
Natural Resource Conservation
  Service, 187
Natural resources
  books and reports, 242,
    243–244, 247
  defined, 286
  scientific journals, 258–259,
    260–261, 262

Natural Resources Canada, 185
Natural Resources Council of
America, 211
Natural Resources Defense
Council, 211
Nature Conservancy, 78, 79, 211
Nature Conservancy of Canada,
211–212
Nature Conservation Society of
Japan, 212
Navajo Indian Nation, 171
Nebraska Forest Service, 194
Net annual growth (defined),
286
Net present value (defined), 286
Netherlands: trade restrictions, 85
Nevada
early forest policy, 7
forest practices programs, 70
private forestry regulation,
63(table), 66(table)
purchase of Federal
timberlands, 119
Nevada Division of Forestry, 194
New Hampshire
concerns over water quality, 71
early private forestry
regulation, 63(table)
New Hampshire Division of
Forests and Lands, 194
New Jersey: local ordinances, 71
New Jersey State Forestry
Service, 194
New Mexico
community forestry in, 59
forest practices programs, 70
private forestry regulation,
63(table), 66(table)
New Mexico Forestry Division,
194
New York
early private forestry
regulation, 63(table)
local ordinances, 71
state Forest Reserves, 120
western reserve claim, 5

New York City
drinking water quality, 96
land conservation, 96
New York State Department of
Environmental
Conservation, 194
New Zealand:
plantations/protected areas
in, 39
Newsweek, 11
Noncommercial species
(defined), 286
Nondeclining even flow, 30–31
Nongovernment conservation
organizations, 197–220
Nongovernmental organizations
(NGOs), 78–79, 97–98
Nonindustrial private forests
(NIPFs), 77, 155–157
area of forest land and timber
land owned, 147(table)
defined, 286
U.S. wood supply, 107
Nonpoint sources, 66–67,
286(defined)
Nontimber forest products, 58
Nontimber forest values
(defined), 286
North American Wetlands
Conservation Council,
185–186
North Carolina
concerns over water quality, 71
western reserve claim, 5
North Carolina Division of
Forest Resources, 194
North Carolina State University,
223
North Cascades Conservation
Council, 212
North Cascades National Park,
13
North Dakota Forest Service, 195
Northcoast Environmental
Center, 212
Northern Arizona University, 223

Northern Pacific Railway, 140
Northern spotted owl, 127, 132,
139, 140, 247
Northwest Ecosystem Alliance,
212–213
Northwest Forest Plan (1994), 92
Northwest Ordinance (1787), 6

O&C Act, 135
Office National des Forêts
(France), 98
Ohio Division of Forestry, 195
Ohio River valley, 144
Ohio State University, 223
Ojai fire (California, 1985), 50
Oklahoma Cooperative
Extension, videos by,
274–275
Oklahoma Department of
Agriculture, 195
Oklahoma State University, 223
Old-growth forest, 30, 133, 135,
139, 286(defined)
Olmsted, Frederick Law, 137
Olympic National Park, 13
Online information
clearinghouses, 265–267
Online reports, maps, and
information, 267–270
Optimization models, 135
Oregon
early forest policy, 7
forest health, 47
private forestry regulation,
63(table), 66(table), 74
purchase of Federal
timberlands, 119
state forest lands, 154
Oregon and California Railroad,
135
Oregon Board of Forestry, 68
Oregon Compromise (1846), 6,
117
Oregon Department of Forestry,
195
Oregon Forest Practices Act
(1971), 66, 67, 68

mandatory guidelines, 68–69
regulatory standards, 69
Oregon State University, 223
video collection of, 273–274
Organic Administration Act
(1891), 8–9
Organic Administration Act
(1897), 16, 89, 90, 120, 148
Organizations, 183–237
conservation, 197–220
state, 190–197
Ownership (defined), 286
Ozarks Resource Center, 213

Pacific Northwest, 30, 271
Pacific Rivers Council, 213
Pack, Charles Lathrop, 137
Parcelization, 156
Partial cutting (defined), 287
Participatory decision-making
processes, 57
Partners in Parks, 213
Penalties, 68
Pennsylvania
forest certification, 171
Forest Reserves, 120
local ordinances, 71
Pennsylvania Bureau of Forestry,
195
Pennsylvania State University, 223
publications and videos by, 275
Periodic annual increment
(defined), 287
*Perspectives on Ecosystem
Management* (video), 273
Peterson, Max, 137–138
Peugot, 98
Pinchot, Gifford, 4, 9, 27, 75–76,
137, 139
biographical sketch, 138
books and reports, 249
as Chief Forester, 120
as "Father of Forestry in the
United States," 138
as martyr for conservation, 121
Second American Forest
Congress (1905), 120

Pinchot Institute for
Conservation, 55, 214, 321
Planning
books and reports, 247
defined, 287
forest, 135
forest management plan, 282
forest stewardship plan, 282
land and resource
management plan, 285
management plan, 285
nonprint resources, 273–274
Plant or habitat diversity
(defined), 287
Plant pests, 103–104
Plant Protection Act, 104
Plymouth Colony, 5
Polar domain, 144
Policy of Forestry for the Nation
(Graves), 134
Ponderosa pine, 48, 49, 159
Powell, John Wesley, 138
Precommercial thinning
(defined), 49, 287
Preemption Act (1841), 118
Prescribed or controlled burn,
126–127, 287(defined)
Preservation (defined), 287
Prior-Appropriations Doctrine,
91
*Private Forest-Land Owners of the
United States* (Birch), 156
Private forest lands, 154–157
books and reports, 240
conserving biological diversity
in, 37–38
cooperative sustained-yield
units, 124
forestry regulation on, 62–74
incidental taking of, 37, 63–64
industrial, 284
management of, 18–20
nonindustrial, 146, 155–157,
286
owners and acres owned,
156–157, 156(table),
157(table), 158(table)

scientific journals, 264
state acquisition of, 127
sustainable management of,
16–20
U.S. wood supply, 106–107
Private forestry regulation,
62–74
comprehensive, 66, 66(table)
early, 63, 63(table)
first generation (1920–1950),
62–65
second generation
(1970–1983), 65–70
third generation (1984–1995),
70–73
Private timber companies, 119
Professional associations,
232–233
Programme for Belize, 98
Progressive Era, 4
Progressive Movement, 9
Prudential Timber Investments,
155
Public Employees for
Environmental
Responsibility, 214
Public forest lands
books and reports, 243, 244,
247
cession to states, 118
conservation of, 75–77
conserving biological diversity
in, 35–37
conveyances to private
corporations, 118
cooperative sustained-yield
units, 124
disposal of, 6–7
early domain, 6
Federal, 147–153
mineral exploration in, 119
original domain, 117
ownership of, 146
timber theft from, 118, 119
total acres, 6
U.S. wood supply, 107
See also National Forest System

Public involvement
    concerns over environmental
        impacts, 72–73
    defined, 287
Public Land Commission, 120
Public Lands Foundation, 214
Puerto Rico, 144
Puerto Rico Forest Service
    Bureau–DNER, 195
Purdue University, 223

Quetico-Superior International
    Peace Park, 123
Quincy Library Group, 55

Railroad land grants, 118
Rainforest Action Network,
    214–215
Rainforest Relief, 215
Rangeland (range) (defined), 287
Recreation opportunity class
    (defined), 287
Recreation use
    indicators of, 167
    of National Forests, 122, 123
    private forest land ownership,
        158(table)
Reforestation, 123
    defined, 288
    performance measures,
        177(table)
    policy encouraging, 122
    standards for, 67–68, 177(table)
Reforestation Tax Credit, 40
Regeneration (defined), 288
Regeneration cut (defined), 288
Regulation(s)
    area regulation, 278
    concept of, 26
    contingent regulation, 71
    L-20 regulations, 42
    on private lands, 62–74
    U-Regulations, 42
Renewable Natural Resources
    Foundation, 215
Renewable resources journals,
    262

Research and development
    books and reports, 250
    capacity to conduct and apply,
        170
    sale of private forest lands for,
        127
    scientific journals, 253
    sustainable development, 32
    water development, 137
Reserva Forestal Malleco, Chile,
    98
Reserve (defined), 288
Reserve mentality, 34
Reserved forests, area, 145
Residual stand (defined), 288
Resources for the Future, 215
Resources Planning Act (1976), 136
Restoration (defined), 288
Revolutionary War, 5
Rhode Island Division of Forest
    Environment, 195
Rio Bravo Conservation and
    Management Area, Belize,
        97–98
Riparian areas, 92
    defined, 288
    nonprint resources, 274
*Riparian Forest Buffers: The Link
    between Land and Water*
    (video), 274
River of No Return Wilderness,
    132
Roadless Area Review and
    Evaluation (RARE), 42–43
Roadless Area Review and
    Evaluation (RARE II), 43, 46
Roadless Area Rule, 44, 160
Roadless areas, 41–46, 123
Roosevelt, Theodore, 9, 75, 137
    biographical sketch, 139
    Second American Forest
        Congress (1905), 120
Rotation (defined), 288
Roundwood (defined), 288
Rulemaking authority, 67
Rural Community Assistance
    Program, 58–59

Sagebrush Rebellion, 76
Sale unit (defined), 288
Salmon, 271
Salmon Forest, The (video), 271
Salvage cut (defined), 288
Sand County Almanac, A
    (Leopold), 135
Sargent, Charles Sprague, 139
Save America's Forests, 216
Sawlog (defined), 288
Sawtimber (defined), 288
Schenck, Carl, 139
Schurz, Carl, 119
Scientific Certification Systems,
    84, 171
    forest certification, 173–174,
    176
Scientific journals, 252–264
Scolel Te Community Forestry
    Project, Mexico, 98
Second American Forest
    Congress (1905), 120
Second-growth forests, 18
    defined, 288
    sustained-yield management
    in, 122
Sediment Control during
    Commercial Timber
    Harvesting Operations
    (West Virginia), 73
Sedimentation (defined), 288
Seed tree cut (defined), 288–289
Seed tree laws, 63
Selection system (defined), 289
Selective cutting (defined), 289
Senate Interior Subcommittee on
    Public Lands, 131–132
Sensitive resource standards, 68
Seventh American Forest
    Congress (1995), 128
Severance taxes, 19, 81
Shelterwood system (defined),
    289
Shipstead-Nolan Act (1930), 123
Sierra Club, 34, 136–137, 216
Sierra Club v. Butz, 43
Silent Spring (Carson), 2

Silvicultural practices, 123
Silviculture (defined), 289
Site class (defined), 289
Site preparation (defined), 289
Slash (defined), 289
Slash disposal and utilization
    performance measures,
    178(table)
    standards for, 178(table)
SmartWood
    categories of chain-of-custody
    certification, 172–173
    categories of source
    certification, 172
    certification for chain of
    custody, 172
    exclusive companies, 173
    forest certification, 171–173,
    176
    nonexclusive companies, 173
    principles of certification, 172
Smokey Bear, 11, 48
Smurfit Stone Container, 78
Social and spiritual needs and
    values, 167
Social ecology journals, 257
Society for Ecological
    Restoration, 216
Society of American Foresters
    (SAF), 138, 233
    accredited forestry schools,
    220–226
Socioeconomic assessment
    books and reports, 241
    long-term benefits, 166–167
Softwood (conifer)
    defined, 289
    trends in, 160
Soil compaction (defined), 289
Soil resources
    conservation and maintenance
    of, 165
    indicators of conservation, 165
Soil type (defined), 289
South Carolina, 5
South Carolina Forestry
    Commission, 195

South Dakota Resource
  Conservation and Forestry,
  196
Southern Arizona Water Rights
  Settlement Act, 140
Southern Hemisphere forests,
  106
Southern Illinois University, 223
Southwest Center for Biological
  Diversity, 216–217
Special sites, protection of
  performance measures,
    178(table)
  standards for, 178(table)
Special-use permits, 122
Species
  defined, 289
  endangered or threatened, 127,
    281, 290
  exotic, 161
  indicator, 284
  indicators of diversity, 163
  keystone, 284–285
  noncommercial, 286
  online tree species range
    maps, 268
  rate of extinctions, 33
Spiritual needs and values, 167
St. Croix International Waterway
  Commission, 186
Stand (defined), 289
Standards, ISO, for sustainable
  forest management, 85
State forest lands, 153–154
State Forest Reserves, 120
State forestry agencies, 88
State of Washington v. Avery
  Dexter, 64–65
State organizations, 190–197
States' Rights Movement, 76
Statewide forest environmental
  impact statement (EIS), 73–74
Stephen F. Austin State
  University, 223–224
Stewardship, 56–57
  books and reports, 240,
    244–245

defined, 289
long-term contracts, 52
nonprint resources, 275
pilot project, 58–59
Stocking (defined), 290
Stora Enso, 78, 99
Strategic Timber Trust, 155
Stream flow (defined), 290
Streamside management zone
  (SMZ) (defined), 290
Streamside restrictions, 67
Strip Mining Reclamation Act,
  140
Strontia Springs Reservoir, 93
Stumpage (defined), 290
Succession (defined), 290
Sulfur dioxide (SO_2) cap-and-
  trade system, 99
Summer cabins, on Federal
  lands, 122
SUNY College of Environmental
  Science and Forestry, 224
Superior National Forest, 123
Sustainability
  forest ecosystem, 24
  of wood sources, 172
Sustainable development, 32
Sustainable forest management,
  24–32
  books and reports, 244–245
  current, 31–32
  defined, 290
  forest principles defining,
    31–32
  guidelines for, 85–86
  legal, institutional, and
    economic framework for,
    168–170
  performance measures,
    177(table)
  on private lands, 16–20
  recent policy developments,
    27–32
  of Reserva Forestal Malleco,
    Chile, 98
  standards for, 85–86, 177(table)
  two-pronged strategy, 35

Sustainable forestry, 4–5, 24
  books and reports, 240
  defined, 4–5
  nonprint resources, 273
  online reports, maps, and
    information, 267
  scientific journals, 259
Sustainable Forestry Board
  (SFB), 175
Sustainable Forestry Initiative
  (SFI) (AF&PA), 32, 86, 87,
    175–176
  objectives or principles,
    175–176
  web site, 175–176
Sustained yield, 24, 124–125,
  290(defined)
Sustained Yield Forest
  Management Act (1944), 124
Sustained-yield management,
  27, 122

Taft, William H., 121
Taking of private land,
  incidental, 37
"Takings" doctrine, 63–64
Tax-delinquent lands, 154
Taxes. *See under name of specific
  tax*
Taylor Grazing Act (1934),
  134–135
Technology
  books and reports, 251
  forest biotechnology, 101–105
Technology Association for the
  Pulp and Paper Industries,
    231–232
Temperate Rainforest, The (video),
  271
Temple-Inland, 99
Tennessee Department of
  Agriculture Division of
  Forestry, 196
Tennessee Valley Authority, 186
Texas A&M University, 224
Texas Forest Service, 196
Texas Purchase (1850), 6, 117

Thaddeus Kosciuszko National
  Memorial, 153
Theodore Roosevelt
  Conservation Alliance, 217
Thinning
  commercial, 279
  defined, 290
  nonprint resources, 274
  precommercial, 287
Thinning Young Stands (video),
  274
Thomas, Jack Ward, 139–140
Threatened or endangered
  species, 127, 281,
  290(defined)
Timber (defined), 290
Timber and Stone Act (1878), 7,
  119
Timber companies, private, 119
Timber Culture Act (1873), 6–7,
  119, 120
Timber famine, 8, 62
Timber harvesting
  annual lumber production, 121
  books and reports, 244
  clearcutting controversy, 15–16
  cut-and-run practice, 17
  Healthy Forests Initiative and,
    52, 129
  import bans on tropical
    timbers, 127
  information clearinghouses,
    267
  long-term sale contracts, 128
  from National Forests, 12,
    13(figure)
  naval reserves, 118
  nondeclining even flow
    constraint, 30–31
  precommercial thinning, 49
  salvage sales, 128
  scientific journals, 262
  standards for, 67
  sustainable volume, 26
Timber investment management
  organizations (TIMOS), 79,
  155

Timber production and trade,
145–146
disputes with Canada, 121
private forest land ownership,
158(table)
Timber theft, 118, 119
Timber trespass, 7
Timbergreen Forestry, 267
Timberland Investment Services,
155
Timberlands
area, 145
transfer to private timber
companies, 119
U.S., 146–147, 147(table), 160
Tongass National Forest, 43, 126,
128
Tongass Timber Reform Act, 140
Total forest land area, U.S.,
158–159
Total maximum daily load
(TMDL) requirements, 92
Tourism, indicators of, 167
Toxic Substances Control Act, 104
Trade
cap-and-trade system, 99
carbon credit trading system,
99, 106, 129
information, 145–146
restrictions on, 85
U.S. disputes with Canada, 121
Trade associations, 229–232
Transfer Act (1905), 75, 120, 139
Trash trees, 62
Tree farm (defined), 290
Tree Farm certification,
177–179(table)
Tree Farm signs, 176
Tree mortality rates, 160
Tree-Sit: The Art of Resistance
(video), 272
Tree-sitting, 272
Tree species range maps, online,
268
Trees Are the Answer (video),
271–272
Trends, 158–161

Tribal forests, 157–158
Tropical deforestation, 127, 128
Tropical forestry
import bans on tropical
timbers, 127
scientific journals, 259
videos, 270
Trust for Public Land, 78, 217
Turpentine Seed Tree Law (1922)
(Louisiana), 63

U-Regulations, 42
UBS Resource Investments, 155
Udall, Morris, 140
UN Environment Programme
(UNEP), 217
Understory (defined), 290
Underwood Tariff Act (1913), 121
Uneven-aged forest (defined),
291
Uneven-aged management
(defined), 277, 291
Uneven-aged stand (defined), 277
Union of Concerned Scientists,
218
Union Pacific Railroad, 7, 118
United Kingdom
Department for International
Development, 98
trade restrictions, 85
See also England
United Nations Conference on
Environment and
Development (UNCED)
(1992) (Rio de Janeiro), 3,
84–85, 110, 128
Agenda 21, 85
"Forest Principles," 31–32, 85
United Nations Economic
Commission for Europe
(UNECE), 269
United Nations Forum on
Forests, 129
United Nations Framework
Convention on Climate
Change (FCCC) (1992), 97,
270

United States
  annual lumber production, 121
  comprehensive forest policy,
    122–123
  consumption of major wood
    products, 34
  current forest policy issues,
    23–116
  current policy developments,
    44–45, 58–61
  directory of forestry
    organizations, 183–237
  emerging forest policy issues,
    106–110
  forest area, 145
  forest conditions and trends,
    158–161
  forest ecoregions, 144–145
  forest ownership, 146–147,
    147(table)
  forest policy, 118–119
  forest policy books and
    reports, 243–247, 248–250
  industry timberlands, 106
  online reports, maps, and
    information, 268
  recent policy developments,
    27–32, 35–40, 50–52, 86–89,
    91–93, 97–100, 103–104
  sulfur dioxide (SO$_2$) cap-and-
    trade system, 99
  sustained-yield forestry in, 27
  timber production and trade,
    145–146
  total forest land area, 158–159
  total wood production, 160
  wood supply, 106
Universal Forest Products Inc., 229
University of Alaska–Fairbanks,
    224
University of Arkansas–
    Monticello, 224
University of California–
    Berkeley, 224
University of Edinburgh, 98
University of Florida, 224
University of Georgia, 224

University of Idaho, 224
University of Illinois, 224–225
University of Kentucky, 225
University of Maine, 225
University of Massachusetts, 225
University of Minnesota, 225
University of Missouri, 225
University of New Hampshire,
    225
University of Tennessee, 225
University of Vermont, 226
University of Washington, 226
University of Wisconsin, 226
University of Wisconsin–
    Madison, 226
University of Wisconsin–Stevens
    Point, 226
*Up in Flames: A History of Fire
    Fighting in the Forest* (video),
    272
Urban ecology journals, 263
Urban forestry journals, 263
Urban-wildland interface
    (defined) 291
Uruguay Round, 84–85
U.S. Constitution, 4, 63–64
U.S. Department of Agriculture
    (USDA)
  Animal and Plant Health
    Inspection Service (APHIS),
    103–104
  Cooperative State Research,
    Education, and Extension
    Service, 186
  Division of Forestry, 120, 134
  Natural Resource
    Conservation Service, 187
  online reports, maps, and
    information, 269
  Rural Community Assistance
    Program, 58–59
  U.S. Forest Service. *See* U.S.
    Forest Service
U.S. Department of Commerce,
    187
U.S. Department of
    Conservation, 135

U.S. Department of Defense, 146

U.S. Department of the Interior, 120, 188–189

U.S. Environmental Protection Agency (EPA), 50, 92, 189
authority over genetically modified tree species, 104
definition of certification, 83
online reports, maps, and information, 268
sulfur dioxide (SO$_2$) cap-and-trade system, 99

U.S. Fish and Wildlife Service (USFWS), 35, 153, 188–189

U.S. Food and Drug Administration, 104

U.S. Forest Service, 11–16, 148, 187, 239
books and reports, 246, 247
Branch of Research, 122, 133
Capper Report, 122
Church Subcommittee Guidelines, 132
clearcutting controversy, 15–16
conserving biological diversity, 36–37
created, 9
decentralized decision-making, 59–60
established as agency of USDA, 120
forest management by, 147, 148
funding, 138
information clearinghouses, 266–267
Inventory and Monitoring Institute, 266–267
land stewardship contracts pilot project, 58–59
long-term stewardship contracts, 52
online reports, maps, and information, 268, 269
ownership of forests, 146, 147
policy of fire suppression, 11, 48

roadless area reviews, 42

Rural Community Assistance Program, 58–59

U.S. Geological Survey (USGS), 189
maps, 6
online reports, maps, and information, 268
surveys, 138

U.S. Navy timber reserves, 118

U.S. Public Interest Research Group, 218

U.S. Timber Conservation Board, 132

U.S. Timberlands, 155

USS *Michigan,* 7, 118

Utah Department of Natural Resources, 196

Utah State University, 226

Vanderbilt, George, 139

Variability, 284

Vermont
concerns over water quality, 71
Department of Forests, Parks, and Recreation, 196
early private forestry regulation, 63(table)
local ordinances, 71

Videos, 270–272

Village of Euclid v. Ambler Realty Company, 64

Virginia
concerns over water quality, 71
Department of Forestry, 196
early private forestry regulation, 63(table)
western reserve claim, 5

Virginia Tech, 226

Volunteer training, 61

Wachovia Timberland Investment Management, 155

Wagner Forest Management Ltd., 155

Washington (state)
Department of Natural
Resources, 196
early forest policy, 7
Forest Conservation Act, 64
forest practices programs, 70
large tracts of low-access
forest, 160
private forestry regulation,
63(table), 66(table)
purchase of Federal
timberlands, 119
state forest lands, 154
State of Washington v. Avery
Dexter, 64–65
Water, 89–93
"first in time, first in right"
allocations, 91
Forests and Water in the Light of
Scientific Investigation (Zon),
141
recent policy developments,
91–93
Water development, 137
Water pollution, 66, 92
Water quality
concerns over, 70–71
defined, 291
forest, 91
nonprint resources, 274
performance measures,
177(table)
standards for, 177(table)
total maximum daily load
(TMDL) requirements, 92
Water Quality Act (1987), 70
Water resources
conservation and maintenance
of, 165
indicators of conservation, 165
Water rights, 91
Watershed (defined), 291
Web sites
information clearinghouses,
265–267
online reports, maps, and
information, 267–270

Weeks Act (1911), 9, 121, 133
Well-managed wood sources,
172
West Virginia
Forestry Division, 196–197
Sediment Control During
Commercial Timber
Harvesting Operations, 73
West Virginia University, 226
Western Governors Association,
51
Western reserve lands, 117
Wetland (defined), 291
Weyerhaeuser, Frederick, 140
Weyerhaeuser, Frederick E., 140
Weyerhaeuser Timber Company,
78, 140
White pine forests, 145, 159
Wilderness, 14–15, 291(defined)
Wilderness Act (1964), 15, 42,
125, 137, 148
father of, 141
Wilderness Act (1978), 42
Wilderness and roadless areas,
41–46, 136, 153
current policy developments,
44–45
designation of, 123, 137–138
eastern, 126
let-burn policies, 50
management of, 153
Roadless Area Review and
Evaluation (RARE), 42–43
Roadless Area Review and
Evaluation (RARE II), 43, 46
Roadless Area Rule, 44, 160
scientific journals, 263
Wilderness Society, 15, 136, 218
Wilderness Study Areas, 126
Wildfire(s), 47
backfire, 278
catastrophic, 159
defined, 291
large-scale, 129
scientific journals, 257
See also Burning
Wildlife (defined), 291

Wildlife Action Inc., 218–219
Wildlife habitat
  performance measures,
    177(table)
  standards for, 177(table)
Wildlife Society, 233
Wilson, Edward O., 33
Wilson Act (1894), 121
Wisconsin
  local ordinances, 71
  total forest land area, 159
Wisconsin DNR–Division of
    Forestry, 197
Wise Use Movement, 76
Withdrawals (defined), 291
WNC Pallet and Forest Products
    Co., 229
Women's Environment and
    Development Organization,
    219
Wood certification, 171
  chain-of-custody process, 174
  principles of, 172
  SmartWood, 171–173
Wood production
  books and reports, 250
  total U.S., 160
  See also Timber production and
    trade
Wood products
  import tariffs and duties on,
    121
  U.S. consumption of, 34
Wood source categories, 172
Wood supply
  global, 106–108
  Southern Hemisphere, 106
Woodland management
    resources, 273–274
Woodlands Resource
    Management Group, 155
Working forest conservation
    easements, 79

Working Group on Criteria and
    Indicators for the
    Conservation and
    Sustainable Management of
    Temperate and Boreal
    Forests, 161
World Forestry Center, 219
World Parks Endowment Inc.,
    219
World Resources Institute, 220,
    269
World Summit on Sustainable
    Development (2002), 57, 110
World Trade Organization
    (WTO), 83
World Wide Fund for Nature,
    85, 109
World Wide Web
  information clearinghouses
    on, 265–267
  reports, maps, and
    information on, 267–270
World Wildlife Fund, U.S
    (WWF), 39, 220
Worldwatch Institute, 220
Wrangell-St. Elias National Park
    and Preserve, 153
Wyoming, 160
Wyoming State Forestry
    Division, 197

Yale School of Forestry, 133, 138
Yale University, 226
Yellowstone National Park, 41,
    126–127, 148
Yield, sustainable, 290
Yield, sustained, 24, 124–125
Yield taxes, 81
Yosemite National Park, 41–42,
    137

Zahniser, Howard, 141
Zon, Raphael, 141

About the Authors

V. Alaric Sample has served as president of the Pinchot Institute for Conservation in Washington, D.C., since 1995. He is a Fellow of the Society of American Foresters and a Research Affiliate on the faculty at the Yale School of Forestry and Environmental Studies. Sample earned his Ph.D. in resource policy and economics from Yale University in 1989. He also holds an MBA and a Master of Forestry degree from Yale and a BS in forest resource management from the University of Montana. His professional experience spans public, private, and nonprofit organizations, including the U.S. Forest Service, Champion International, The Wilderness Society, and the Prince of Thurn und Taxis in Bavaria, Germany. He specialized in resource economics and National Forest policy as a Senior Fellow at the Conservation Foundation in Washington, D.C., and later as Vice President for Research at the American Forestry Association. Sample has served on numerous national task forces and commissions including the President's Commission on Environmental Quality task force on biodiversity on private lands and as cochair of the National Commission on Science for Sustainable Forestry. His other professional activities include serving as chair of the National Capital Society of American Foresters and chair of the board of directors for the Forest Stewards Guild.

Antony S. Cheng is on the faculty in the Department of Forest, Rangeland, and Watershed Stewardship at Colorado State University in Fort Collins, Colorado, where he teaches courses in natural resource history, policy, and sustainability. Dr. Cheng's research and practice focus on improving the quality and durability

321

of natural resource decisions through more meaningful public participation, collaborative approaches, and institutional innovations. His work is published in the *Journal of Forestry, Society, and Natural Resources* and *Forest Science.* He received his Ph.D. at Oregon State University in Corvallis and MS at the University of Minnesota in St. Paul, both in forest resource policy. He holds a BA in politics from Whitman College in Walla Walla, Washington.